DRINK FROM THE BUCKET

Adventures and Introspections from a Feral Life

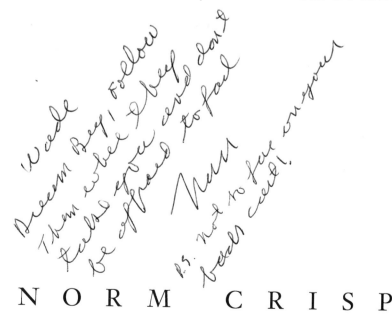

NORM CRISP

© 2017 Norm Crisp
All rights reserved.

ISBN: 1978282389
ISBN 13: 9781978282384

TABLE OF CONTENTS

Preface	v
Foreword: Preston Larimer	vii
Trout by a Hair	1
Skunked	13
Final Loss of Innocence	19
Norm, Not Norman	26
It's Genetic	30
New Zealand – Julia the Witch, The First Adventure	34
Pakistan Customs	39
Marmot - It's What's for Dinner!	48
Running With the Bulls – Life Lessons	53
Chasing the Sun - Follow Your Quest	59
Traveler or Tourist – The Himba People	64
Maturing Streams—Growing Older	68

PREFACE

As a boy growing up in a small town in New Hampshire I never dreamed my life would turn out the way it has. Life has been damn good to me. Yes, there have been a few hard spots, but nothing that has stopped me; life just made me take a detour or two. Along the journey I've had some interesting adventures and moments of reflection.

Over the years, friends and acquaintances have told me I should write a book about my experiences. I never really thought of my adventures as book-worthy; they were just how I spent my life.

About a year ago I was rummaging through my collection of travel journals. It occurred to me I might not be around to share my stories and introspections with my potential grandchildren. I needed to find a way to share my life with them. My unborn grandchildren were the inspiration for this endeavor.

FOREWORD

I became a trout fisherman many years ago on a wild brook trout stream in Central Pennsylvania. The stream was small enough that I could jump across any point without getting my feet wet. I took a small spinning rod, a worm hook and some red wigglers, and a red and white plastic bobber, and cast the worm through the hemlocks and mountain laurel. Somehow I landed the whole rig in a tiny pool. I wasn't prepared for the little brookie that grabbed that worm. It wasn't its small size or my luck as a fisherman that surprised me, but its beauty. The pure loveliness of that animal has stuck in my head ever since. A few weeks later I picked up my first fly rod. At that moment, I started down a trajectory of life that eventually brought me together with Norm Crisp.

There have been thousands of trout in my life since. That first brookie was 40 years ago, and I've spent a lot of time standing in moving and still waters ever since, including 20 years as a fly fishing guide. Many of those trout have been forgotten, but there are many I remember. I can recall not just the fish themselves, but also the fly they took, the location (down to the pocket of water where they lived), the presentation of the fly, and the ecological conditions that caused that fish to come to the fly. As I think back, the trout I remember the best are often ones I shared with Norm. I say share because we both remember each in the same way. Today, Norm and I don't see each other as much as we did over the 25 years we spent each fall, winter, and spring fishing in the Missouri Ozarks, and each summer on the Encampment River and its tributaries in Southern Wyoming. Norm will say to me, "Do you remember that nice brown you caught just above that irrigation diversion at the edge of the Encampment River Wilderness in Purgatory Gulch? He came out of that log strainer and took that size 10 yellow Stimulator." Then, he will proceed to tell me every detail of the catch, often better than I could remember, including reminding me that it measured 16 ¾ inches, not 17. We shared the memories of catching (or often missing) those fish, and that is the type of friendship we have always had. Sometimes he out fishes me, sometimes I out fish him, but we can walk up the edge of a river, casting our flies so close together they would land mere inches apart. All the while, we would discuss life, ecology, raising our sons, and jobs (I taught environmental science and he was a water scientist). At the same time, we would always be a few teasing jabs in about the last cast or whatever. Few people could fish that way, with another person so close by, but it has always worked for Norm and me.

I thought I knew everything about Norm, having spent countless hours together on various streams, in cafes, and around the campfire. But I learned that wasn't true. In reading this book, I see there was much still to learn. We have our stories too; lots of them, but those will have to wait for another book. Many of our stories took place in the same places that are in this book, like Hog Park Creek, or in the BLM campground near the town of Encampment. That campground is on the river next to some summer homes owned by members of the Odd Fellows Lodge in Rawlins, WY. Thus, we've always called that campground "The Odd Fellows Campground". There we met many of the enduring characters of our lives- Charlie, Georgie, Sarah, James and Annette, Duane and Connie, Steve, Travis, Denver, Pete and the deaf mushroom hunters. Then there were the prairie dog hunting brothers, Matthew, Mark, Luke, John and Jerry, one of whom once told me that "you always bring something on a camping trip that you don't need" as he watched my wife prepare dinner in our campsite. My son has urged me to get together with Norm and write down these stories someday. He feels we should just entitle them "Odd Fellows". He just might have it there, a life with so many odd fellows, brought together by the prospect of stalking trout. That certainly is part of the bond that brought Norm and me together almost 25 years ago, and it keeps us going today. But more than that, it is Norm living life outside of the norm, and what you are about to read is what living outside of the norm can be, taking that road less travelled.

Norm Crisp ponders his life in this book. Recently, I saw a cartoon with a coffin and a tombstone with the epitaph "You have the rest of eternity to think inside the box." I thought of Norm when I read it, as he has always traveled while thinking outside of the box. Years ago, he told me he felt that he had earned the right to be a curmudgeon. But along the road to curmudgeonry, I think he missed the mark and instead became somewhat of a sage. Never does he stop thinking of life, experiences, friendship, the ecology and history of fly fishing, and where his life has taken him. What follows here is a sampling of those stories and his wanderings that led him into a good life.

Preston Larimer
Buena Vista, Colorado

TROUT BY A HAIR

My buddy Charlie and I met some years ago while camping and fishing on the Encampment River in Southern Wyoming. I don't know exactly why, but we quickly became the kind of "best friends" who just met. We corresponded by letter throughout the year, filling each other in on our fishing exploits, and met again on the Encampment the following July. We've maintained the connection year after year, letter after letter, always cementing our friendship with an annual pilgrimage to the Encampment River.

    Charlie always got the best of me in our exchange of fish tale letters, but that never stopped me from trying to best him. One June I took a trip to Chicago, not far from where Charlie lives in Wisconsin. It was late spring, about four years after Charlie and I had first met. In our letters we always talked about fishing together in Wisconsin, but life always got in the way. I was so close, I decided to visit; it was an opportunity to spend an evening bragging of fishing exploits and fishing with him on one of his favorite local streams, Black Earth Creek. As we walked back to the car after a great evening of fishing, Charlie spotted some horse hairs on a pasture fence and paused to pick them off. With a twinkle in his eye he asked me, "You think you could land a trout on one of these? That's how old Izaak Walton had to do it." Our good-natured rivalry wouldn't let me pass up the challenge. I quipped, "Not only could I catch a trout, but I could catch a big trout." Looking back I can see how easily I rose to take his bait. After a great deal of discussion over the details of the dare, we settled on 15 inches or more as an acceptable definition of "big".

Back at home with the horse hairs Charlie had found, I started to have second thoughts. I tugged on one of the horse hairs; it snapped easily, telling me even though my horse hair had a diameter of about .008 inches, they weren't as strong as typical 3X to 4X tippet material, which can hold seven or eight pounds. I decided now was a good time to start re-reading my copy of *The Complete Angler*, a Christmas gift received from a friend a few years earlier. I found the information I needed in chapter 21, titled, "Direction for the making of a Line, and for the coloring of both Rod and Line". According to author Izaak Walton, hairs from a light colored horse were the best if you could find one that wasn't flat and uneven. However, Walton cautioned they were difficult to find. He wrote, "If you get a lock of right, round, clear, glass-color hair, make much of it." The second best hairs, he pronounced, are the black hairs. Luckily, one of my co-workers had horses. I asked him if I could have some of their tail hairs, and though he looked at me rather strangely, he invited me to pluck all I wanted. A Saturday in the country gave me a lot of potential tippets.

Back home, with the dining room table cleared of its usual clutter of unopened mail, folded laundry waiting to be put away, and a dirty dish or three, I started examining my booty in search of a "glass-color hair that I could make much of". As I scrutinized each one, I found Walton was right. The donor horse had a very sparse tail, and I didn't harvest a single glass-colored tippet. They were all flat and uneven, and broke with the slightest pressure. I decided to forge ahead with what I had; I'd always liked black tippets anyway!

Armed with a sandwich bag full of black horse hairs, my weight rod, and some "sponge spiders", I headed for my favorite local farm pond for some experimentation on bluegills. It was early June, and the bluegills were on the spawning beds. Finding and casting to bluegills would not be the issue. It was tornado season in Kansas, the afternoon typically sultry: hot and humid with thunderheads building to the west. I had no doubt about my ability to put my fly in the right place to entice a bluegill. My doubts surrounded whether I would overreact and set the hook too hard, whether the horse hair would be strong enough, or both? Would the horse hair be strong enough?

On my third cast, a palm-sized bluegill inhaled the spider, along with half of my hair tippet, before taking off. That was my first lesson. I hadn't checked the black hairs for, as Walton put it, "galls and scabbiness". Close examination of the other hairs revealed they all had these areas somewhere along the length of the hair. I could find it by grasping each end of the hair between my thumbs and forefingers and giving a quick jerk or two. The hair would inevitably break at the scab, leaving me with a 15 to 18-inch length of usable tippet. I also learned horsehair is brittle, and that makes it hard to tie knots that don't break when tightened. I headed home after dark that night a little disappointed. I hadn't caught any fish, and it looked like a "big trout" might be out of the question.

About a week after the bluegill experiment, the answer to the brittle hair problem revealed itself to me as I took my morning shower. Old Walton had said, "First let your hair be clean washed." He had been right about light-colored hairs and scabbiness, so why wouldn't he be right about cleanliness? I decided to give it a try that evening. I shampooed the hairs and soaked them in conditioner, praying all the while that none of my friends would call and invite me over for dinner or to go out to a show. They never would have believed me when I gave them that ludicrous excuse we've all heard, "I'd love to, but I have to wash my hair." In this case, my horsehair tippets. This "salon treatment" really helped soften the hairs, but they parted at the knots frequently. After attempting several combinations I settled on a surgeon's knot for the leader-to-hair connection, and a loose Duncan's Loop for the hair-to-fly connection.

Armed with information about the best knots and 20 of my finest, hand-picked and shampooed tippets, I was prepared to head for the Encampment River for a week of friendship and fishing with Charlie. I could taste the possibility of catching a trout as a "Complete Angler" would.

It is an 800 mile drive from the Kansas City area to the Bureau of Land Management campground on the Encampment River where Charlie and I rendezvous. Even with an early morning departure and gaining an hour crossing into Mountain Time Zone, it was early evening before my little truck and I finally got to rest. Renewing friendships and setting up my portion of camp were the pressing business. When the work was done, Charlie and I settled in to inspect my tippet selection and talk fishing as we unwound from our respective journeys. Fishing would have to wait until morning.

The trout in our preferred spot on the Encampment River are very civilized. It's generally at least 8:30 in the morning before they even consider rising to the best presentation. By then, the sun has started to clear the canyon rim. Though some might complain about the late start, Charlie and I saw it as a "social grace", allowing us plenty of time to drink coffee and prepare for the day's fishing. With all the fanfare I could muster, I rigged out my rod and ceremoniously chose my finest tippet. Besides being the best fisherman I know, Charlie is a good fly tier. It only seemed fitting that one of Charlie's flies should adorn the end of my horsehair. We started the selection process by checking the rain flies on our tents for newly-emerged insects. After a taking a quick inventory of our resources, we decided on a starting pattern and determined a dark brown mottled caddis, about size 16, might be the right choice to match the insects. Charlie's "Woodchuck Caddis" would make a good match. As I finished my coffee, I asked Charlie if I could have a "woodchuck". With a gleam in his eye he handed me one and warned me, "Remember, this one always gets rises from the biggest trout on this river. I'm not sure you could land anything over 13-and-a-half inches, no matter what strength tippet you use."

Charlie and I have a very similar style of fishing. We only work the most productive-looking areas, and we only give them a few drifts of our best possible presentations before we move on. This way we get good coverage of likely lies and spend less time staring at marginal waters. This leads to more time drinking in the sights, sounds, and smells of the river. We often fish together, each taking a side of the river. On this day, a coin flip awarded me the left side of the section we were going to fish; the side with the best holes. On the first series of passes, Charlie connected with a brown of about 12 inches. The day was starting out right. There was nothing for me in the first pool, but on the first drift across the second hole, a brown made a wild splashing rise.

Trying to balance my strike with a force that would set the hook without breaking the brown off at either of the leader knots, I raised my rod tip. The knots held, and battle commenced with my first "Horse Hair Trout". I was so afraid the knots might fail, I played him like he was that one big fish that comes along during each trip. (You know, the "oh shit" fish.) Slowly, I worked him to me and gently slid the net under him. I had caught a trout using a horsehair. As nonchalantly as I possibly could, I held my 11-inch treasure up for Charlie to see before I slipped him back into the river. With a smile, Charlie reminded me that we had some good water ahead of us where I could catch a few more practice trout before we reached the big rocks, a place where big trout often hang out. I started feeling confident I could land a big trout on a hair.

As we worked our way up the river, talking about life in general and what it was about trout fishing that raises our passion, we caught several more trout. None were big, but each of these practice trout told me as long as I didn't try to "horse" them in, the hair would hold. Of course, the decision wasn't entirely mine. Trout can be categorized into two broad classes based on the way they rise. Average trout, those somewhere in the nine to 13 inch range, tend to be rather emphatic about how they go for a fly. They generally make a spectacle out of the take, with a kind of wild, splashy rise. It reminds me of the way my sons Ethan and Dan go for the last cookie in the jar if they think the other one wants it. The other type of rise is what I call the "minnow" rise. This is the tricky one. It is just a gentle nudge of the fly. It could be a little creek chub that is having trouble getting its mouth around a size 14. Then again, it could be one of those trout that can inhale a dun from two feet away.

We came to a series of big rocks, which indicated a change to the river's slope. Upstream, there were a series of long pools. Downstream, the river was composed of rocky riffles with pockets. The trout habitat at this juncture was a combination of the best of the upstream pools and the downstream riffles. Big rocks often equal big trout.

As we approached, Charlie headed for the bank and found what he must have considered the prime vantage point from which to watch me. The earlier coin flip had been for effect. I was being awarded the entire width. There were so many current seams and holes, it was hard to decide where to start casting. I spied one hole that looked particularly productive, and decided to start there. As the current came around a rock it formed a little backwater not much bigger than a skillet-sized pancake.

Whiffs of foam moved along the current seam and collected against the backside of the rock along with any insects fortunate enough to run the gauntlet. This hole belonged to the family of one-cast pools. If there was a trout in the pool it would probably rise to my first offering. The question was, would it be a "cookie jar" rise or a "minnow" rise? I dropped my fly right on the line that separates the main flow of the current from the backwater. It had no sooner touched down then I got a "minnow" rise. My heart pounded with excitement. There aren't any creek chubs in the Encampment River. About three inches of trout head rose out of the water, then dropped to the bottom. A quick lift of the rod tip didn't do a thing. It was as if a trout-colored log had my fly. Suddenly, my line moved a few inches forward only to stop and float back at me. Without the slightest effort the trout had parted my hair at the knot connecting the tippet to the leader. I waded over to the bank where Charlie sat. He gently nodded his head up and down and said, "Nice one, wasn't he." Walking back to camp, Charlie pulled a fly box from his vest, took out another "woodchuck", and handed it to me without saying a word.

One of the Encampment's main tributaries is called Hog Park Creek. Hog Park is a broad meadow nestled about 9,500 feet up in the Snowy Mountains. Hog Park Creek is a nice little ten to twelve-foot-wide headwater stream that meanders through the valley. It must not be in any hurry to make the four or five mile trip to the Encampment because it keeps looping back on itself, as meadow streams have a way of doing. With all the bends and their undercut banks, flow-stabilizing beaver dams, and the habitat improvements made by the U.S. Forest Service, the odds of connecting with a "big" trout are very good. Even if the odds weren't good, it is so pretty it would be worth the trip just to practice casting. We decided to make the 15 mile excursion up the mountain along the Forest Service Road, a twisting and turning road made more treacherous by wash boarding.

About halfway across the meadow, Hog Park Creek takes a fancy for one side of the valley and flows along its edge. At the spot where it first meets the hillside, the creek makes a 90 degree bend instead of one of those lazy, meandering turns. Here, the bend has formed a deep pool with a long undercut bank. It's the kind of spot where a big brown might just decide to retire after moving up from the river on its fall spawning run. It is also next to impossible to fish, unless, of course, you are a left-handed caster with a great sidearm, reach, and double haul cast, and can throw a nice "S" in your leader. If you can position yourself close enough to the hillside bank, and the wind is right, you can place your offering right where the creek drops off the gravel bar and starts toward the bend. The depth varies anywhere from a foot, to five feet, before the creek takes the right angle turn and the undercut bank begins. If, and that is the operative word, you can manage to make an extraordinary cast, it is almost guaranteed you will draw a rise. Sometimes it is the "cookie jar" variety, but most often it is the "minnow" type.

About this time, Charlie decided he preferred brookies over bacon to go with our eggs the next morning. He headed for the nearest beaver pond to catch some while I continued to fish my way up the river. As he left me to my question he hollered over his shoulder, "You may be a fair big trout catcher, but we both know your casting is a little suspect. I know exactly where you are heading, you better hope that not only did the Gods of the Rods bless you, but the wind is blowing in the right direction as well." I set to my task with his words of "encouragement" echoing through my head.

My knots were holding, I think in part because I had started greasing the hairs with float paste before tying and tightening the knots. This provided a lubricant that must have reduced damage from friction. My hair held for several sub-big browns and a couple of breakfast-sized brookies. But by the time I reached the bend, it was obvious the Rod Gods were not on my side. Moving in as close as I could to the hillside bank without overtopping my chest waders, I made my first cast. It fell short and wide of the mark by more than I care to admit. Thinking more about the shortcomings of my cast than the dragging drift of my woodchuck caddis, I wasn't prepared for the assault on the cookie jar. By the time I realized that some dumb trout had gone for my botched presentation, it was futile to strike, and he was gone. Figuring my first cast was so far off the mark that Mr. Big couldn't have been spooked, I took a deep breath, mustered all my skills, and made a second cast. It was almost an instant replay of the first cast, right down to the cookie jar rise. There were, however, two differences: first, I was prepared for the take; second, the trout hit so hard he knocked the cookie jar off the shelf. In one fluid motion, he took the fly and headed straight up the creek and over the gravel bar with half his back out of the water. He stopped for a second to "catch his breath" in the next upstream pool, then turned back over the gravel bar toward the undercut bank at the bend. Stripping in line as fast as I could, I managed to just keep him from his safe haven. Trying to negotiate to a shallower and better position, I lost a few inches of the line to him. This was all he needed. He immediately took advantage of this golden opportunity to practice his tying of tippets to roots. Of course by this time, Charlie had caught all the brookies we needed for breakfast and witnessed the entire episode. He informed me that one of the finer things associated with trout fishing is, "the opportunity to see a dumb fly fisherman and a smart trout match wits." I'd had two chances for a big trout on a horsehair and had tallied two misses. We called it a day and headed back to camp.

    I started the next morning fortified by a breakfast of brookies pan-fried in a little wild sage, eggs over easy, and a cup of strong camp coffee, warming myself in the first rays of sunlight as they topped the rim of the canyon. After finishing my coffee I hiked about five miles up the river to a spot near the junction of Hog Park Creek and the Encampment River. Though it's not far, there is so much good fishing water closer to camp that not many people bother to make the trip that far up the river. The area I wanted to fish is canyon country. In this section, the Encampment rushes through a steep walled flume. In a few places the flume gives way to less rugged conditions on one bank or the other. In these areas the velocity slows a hair, and the river gets a little bit tamer. One oasis in particular has always held a big trout for me during trips I have made in October in search of spawners. I always get a rise, but I don't always land the trout.

By the time I reached the area I wanted to fish it was nearing noon and the sun was at its fullest. Nothing seemed to be emerging and I didn't see any rises. With the angle of the sun it would be hard to see my fly on the water, even with polarized sunglasses. Since nothing seemed to be happening, I figured I might as well use something like a size 14 Royal Wulff that I could see fairly easily. Experience with my lucky pool told me my best chances were at one of two spots. The first is about halfway up the pool near a boulder that rests just below the surface. The second is in the eddy that forms where the tongue of fast water races past the ledge on the left side of the river. Slowly working my way upstream, I covered the boulder area from every angle. Each cast floated back toward me without stirring the interest of a single fish. It looked like the tongue was my last opportunity. Stopping just short of the best casting position, I tested my knots and gave my fly and horsehair a fresh layer of grease. I scanned the water for a telltale flash or movement that would reveal a trout's position. Though I cupped my hands to the side of my face to eliminate as much glare as possible, other elements worked against me. The water was too deep and the surface too choppy and broken for me to see anything.

Just as it had been at the boulder, every drift of my fly passed through the tongue and over a potential lie as if it was barren. I grew discouraged and hungry, the call of the squashed peanut butter and jelly sandwich in my vest pocket like a siren's song. I decided to take one more cast, then eat my sandwich. I let instinct take over, giving no thought to where I was casting. My Royal Wulff landed in the very heart of the current where the water ran the fastest. Out of the corner of my eye I caught a glimpse of motion as my big trout shot up from the bottom like an arrow toward the fly.

Without even rippling the surface, he took the fly and plunged to the bottom with such authority that he set the hook for me. Once at the bottom, he turned and rocketed back to the surface. He broke through and reversed direction, then sliced back into the water like a springboard diver. In what seemed like slow motion, he repeated his performance for a second time, then a third. Each time I was certain I'd seen the last of him. You could almost feel his anger and frustration at not being able to throw the fly. With a burst of energy he used the current to his advantage as he made a run for the mid-stream boulder. Afraid to put too much strain on the horsehair, I lowered my rod tip and pointed it at him. Pulling my line to the start of the backing, he reached the boulder and shot into its shelter.

I knew I had to get him out from behind the rock before the chaffing wore through the tippet or the leader. Reaching out across the current with my rod parallel to the water, I was able to apply enough lateral pressure to lead him out from behind the boulder. Though he still had some fight left in him, it seemed that evicting him from behind his rock had broken his spirit. As quickly as I dared, I moved him toward me in the slower and shallower water along the bank and into my net. For a moment I just stood there and looked at him. The realization that I had in fact caught a "big" trout on a horsehair suddenly hit, and I spontaneously started doing a little jig that my sons call "Dad's happy feet", shouting, "I did it! I did it!"

If catching him had been hard, deciding what to do with him was even harder. I generally release most of the fish I catch. I ate so many trout when I was a kid growing up in New Hampshire that I don't have much of a taste for them anymore. Charlie would have kidded me a little, but he would have believed me if came back to camp and told him I'd caught my "big trout" and released him. This trout had fought like hell, a very worthy adversary. I don't know if it was pride in my trophy or the primordial instincts of the hunter as provider (probably a lot of both), but after an internal struggle, I decided to bring him back to camp.

I didn't fish anymore, and by now my peanut butter sandwich was completely forgotten. Carefully taking my prize from my net, I broke off the horsehair tippet from the leader, leaving the fly and tippet mated to my trout. There is a fine line between pride and arrogance, and I felt both. I was proud of my fly fishing skills in knowing where to place my fly for the best fish, and I was arrogant knowing I had just accomplished something few trout fisherman had done in a very long time. Of course, the joy of being able to lord it over Charlie was there as well. I set to my task of bringing my catch home. As a boy in New Hampshire I had wrapped the fish I caught for dinner that night in moistened grass or ferns to keep them fresh. Plucking some grass from the stream bank I ceremoniously wrapped my treasure, as I had done in my youth, and placed it in the back pouch of my vest for the trip back to camp.

The walk back down the canyon didn't take nearly as long as the trip up had taken. Certainly the downhill grade helped, but there was admittedly an extra bounce in my step and an ear-to-ear grin propelling me along. As soon as Charlie saw me he shook his head and said, "You did it, didn't you? I knew you would sooner or later. It was just a matter of time. Well, let's see this treasure of yours." Slipping out of my vest, I removed my prize and laid it out for Charlie to see. After admiring it and teasing me about leaving the fly and tippet in its jaw, he said, "Well, to make it official I better get a tape out and measure him." I think Charlie always measures trout a little short, at least when he is measuring mine, so when he pronounced my prize as "officially big" at 16 inches, I knew I had easily, at least as far as size goes, met his challenge. Halfway through my recount of the catch, Charlie excused himself, rose from the log he was sitting on and headed for his tent. He returned with an old blue book in his hand. As I finished my story, Charlie was thumbing through his book. Having apparently found what he was looking for, he looked up at me and said, "Well you did it. But you did it the hard way. Your copy of *The Complete Angler* must not be the Izaak Walton and Charles Cotton edition. In that edition, 'The Second Day, Chapter V - Of Fly Fishing' says, 'But he that cannot kill a trout of 20 inches long with two deserves not the name of an

Angler.' Walton and Cotton twisted two hairs together for a tippet. They didn't use a single hair like you did. I think a 16-incher on a single hair is equal to a 20-incher on two."

As the sun started to make its way downward toward the sage-covered hills to the west of camp, I started thinking about fixing my treasure for a shared victory dinner with Charlie. When the sage wood fire had died down to a bed of perfect cooking coals I gently placed my prize, its body cavity filled with a few sprigs of sage and slivered garlic, on the grill. As I turned it over I called out to Charlie to bring his plate to share my bounty. He said he would be right over as soon as he found something he had in his tent.

Having found what he was looking for he walked over to the fire. With a glimmer in his eye he handed me a tattered book and said, "Hey Angler, why don't you take my copy of *The Complete Angler* for future reference."

SKUNKED

I'm sure every language has them, but it seems American English, compared to British English, has more than its share of descriptive and colorful words. Maybe the hybridization of so many immigrant languages has created the patchwork of our verbal quilt. Those of us with American English as our native tongue take these words and phrases for granted. I know I always have. But they creep up fairly often as I speak with non-native English speakers. I do a fair amount of international traveling to fish and give programs and seminars. Most of my European comrades and audiences speak English, and a few are even fluent in American lingo. But inevitably, we run across an American phrase that has no English equivalent. "Skunked" is a perfect example.

"Skunked" has been a part of my vocabulary for as long as I can remember. Did you ever remember not knowing what "winter" meant? If you're a fisherman, the same goes for skunked. You know exactly what it means: You went fishing and didn't catch anything. You got skunked.

A couple of years ago I was pike fishing in the Netherlands with my friend Gerlof. I had been casting big streamers all day in a canal that drains a residential area of Zwolle. It's one of Gerlof's favorites because he loves hooking a big pike in front of residents having their morning coffee. On this particular day, Gerlof had caught two, but I hadn't had a strike or a follow all day. It was getting late, and I was getting discouraged. Reluctantly, I asked Gerlof for five more minutes before we called it quits. Then it happened.

The strip of my streamer stopped dead. It was as if I had snagged an underwater log. Then my rod nearly bent in half as a canal monster dove for the bottom. I caught the pike of the trip, a beautiful, 40-inch river wolf, with just two minutes to go!

As we released it, I told Gerlof, "Damn, I was sure I was going to get skunked!"

With a quizzical look on his face and his head tilted to the side, he said, "Why skunked?"

All I could think of to say was, "Why not skunked?" Not the best of answers. On the flight home from Amsterdam I rolled "why skunked" around in my head. Then I remembered…

I went to college in the hills and hollows of eastern Kentucky. Things in that part of the world are different from where I grew up in New Hampshire. First off, they speak a different dialect of English. And second, coon hunting is more important there than fly fishing! In college, my part-time job as a day laborer for a local homebuilder gave me the opportunity to meet and work with Claude, the owner of one of the most renowned coon hounds of the area at that time.

Coon hunting is different than just about any other kind of hunting. If you're not familiar with how it works, it goes something like this: After working all day, you go out in the woods in the middle of the night and run around chasing barking dogs. And coon season is during the winter, so you can freeze your ass off as you are having this great time! I hadn't been much of a hunter back in New Hampshire. In the fall we would use my rowboat to chase ducks around Bradley Lake, but that was the extent of my hunting pursuit.

Working for and listening to Claude for a few semesters while I carried 2 x 4 lumber or pushed wheelbarrows of cement, I learned a few things about coon hunting. First, a coon hunter's status and reputation is, to a great extent, a reflection of the quality and abilities of his coon hound(s). Second, no better word describes these dogs than "hound". They are lean, lanky, and strong, with big sad eyes and long floppy ears. They somehow remind me of a fit, lanky freshman from some remote town, who just arrived on scholarship at state university to play tight end on the football team. Both exude raw energy, talent, and strength with a pure innocence and passion for the game. The only difference is the football player generally has smaller ears!

Claude's hound was "Ol' Major." Ol' Major's lineage was well-known, but his pedigree was another issue. Claude thought he was mostly Blue tick, but he suspected a Treeing Walker might have slipped in sometime.

Third, being the local champions, Claude and Ol' Major had some social responsibility to mentor aspiring coon hunters and their hounds in the fine points of running a coon. This mentoring was accomplished by letting a newcomer hound, which, I learned, was called the "Newbie," and Jim, his owner, the cousin of Claude's wife, hunt with Ol' Major and Claude. My initiation to this sport took place as a ride-along on one such mentoring hunt.

One Monday afternoon in early December, Claude offhandedly said if I didn't have too much studying to do that night, I was welcome to be part of a hunt. Being an outsider, especially from New Hampshire, it was an honor and privilege—and loving the outdoors, I was always up for a new adventure. How could I pass it up? Organic chemistry be damned, I was going. All I needed to do was be ready at 8 pm with a light, warm clothes, and a comfortable pair of boots.

Just moments before our meeting time, Claude's pickup truck arrived. I was out of the house in a flash. Ol' Major and The Newbie were chained in the bed of the truck. Jim slid out of the truck so I could get to the middle of the cab and saddle myself around the gear shift on the floor. It wasn't the best seat, but I was the new guy, after all.

The hollow we were going to hunt that night was on a branch of Brownie's Creek in Whitley County, not far from the Kentucky-Tennessee border. To get there, we endured a torturous half-hour of bumps, turns, low-water crossings, and steep hills through the deep woods. I squirmed continually to avoid the gearshift between my legs. This old road only saw coal trucks during the day and moonshiners and the occasional coon hunter at night. I could only imagine what The Newbie was thinking of this bumpy ride in the back, the newbie in the front wasn't having a much better ride!

Finally, we reached our destination. Claude parked the truck in front of a long-abandoned schoolhouse, eerily lit in the dim glow of a first-quarter moon. Claude and Jim quickly hopped out of the truck to release their dogs while I untangled myself from the gearshift. As soon as the dogs were out of the truck bed, they ritualistically circled the truck, stopping to do their business in turn at each tire. Newbie instinctively knew his place in the pecking order and stayed a respectful distance behind Ol' Major. And after each stop, Ol' Major gave him a slightly menacing glance as he put his hind foot back on the ground.

From the truck, Claude retrieved an old Army surplus knapsack, a blood-stained and tattered grain sack, and an old single-shot .22. Claude kept the knapsack, Jim got the gun, and I got the grain sack. Then, with the words, "Go run one ta' ground Ol' Major," the dogs were off in a flash, Ol' Major in front and The Newbie right behind him.

Anticipating my inexperience, Claude told me we should just wait there on the tailgate until Ol' Major "opened up"—coon hunter speak for the barking howling bawl a hound emits when it starts to track, that is, "gets onto" a live scent. As the silence of these thick winter woods settled around us, Claude never doubted for a second that Ol' Major would open up—it was just how soon and in which direction. After about 10 minutes of wishing I had worn another layer of clothing, the haunting and magical sound of Ol' Major's distinctive croon began drifting down Brownie's Creek like a fog horn. Within seconds, we were off into the woods following the direction of Ol' Major's bawling.

Another thing about coon hunting I learned that night is that there is honor among coon hounds. An instinctual protocol dictates that only the first hound to get onto the track is permitted to bawl. Apparently, the other dog just runs along for the sheer excitement of it. Ol' Major, being the champion coon hound, picked up every track, fresh or old.

Although a coon hunter is always proud of his dog when it "gets to runnin'", following an old track is hard work in the form of a long, rapid hike over and around every obstacle in the hollow. We were on an old track. After what seemed like an hour of sliding down every bank and jumping down every log in the watershed, the pace of the bawling suddenly increased. This crescendo elicited a fatherly pride as Claude announced that Ol' Major had "run the coon to ground".

Our pace increased to match that of Ol' Major's bawling as we neared the tree in which the coon was trapped. From a respectful distance, The Newbie circled the tree with a quivering gait while Ol' Major made one vertical leap after another up the foot of the tree, bawling excitedly all the while. Claude pointed the beam of his headlamp into the tree's branches and searched for the telltale red glow of the coon's eyes. No eyes appeared.

After a good while in unfulfilled anticipation, Cousin Jim started to announce, with a touch of sarcastic glee, that Ol' Major, The Wonder Hound, had been fooled. Just then, however, Claude caught the faint glimmer of red eyes from a nest hole partially obscured by a limb. Jim immediately began to praise Ol' Major's prowess in running such an old track to fruition. Meanwhile, Ol' Major was still trying to climb the tree to get to the coon. Claude walked over beside the agitated dog and simply whispered, "Good boy, Ol' Major." Like magic, Ol' Major stopped trying to scale the tree. Exhausted, he sat down on his haunches to catch his breath. Then Claude turned to us and said, "Coon's in a hole."

Claude looked around for a suitable log for us to sit on. Once he found one that would hold all three of us, he opened the knapsack and produced a snack that, in our cold and expended state, was fit for a king: a box of soda crackers, a sweet onion, and a can of potted meat. Then Claude passed around a mason jar of what looked like water. I had to catch my breath after my first stinging drink. I never knew a first quarter moon could shine so brightly!

Refreshed from our snack and short rest on the log, Claude said the magic words again—"Go run one to ground, Ol' Major." And with that, both hounds disappeared into the dark again. This time we didn't have to wait very long for the symphony to fill our ears. In just seconds, the tenor bawling of The Newbie rose from the still woods. Jim beamed like a father at a child's first recital. His dog had found the track, and found it immediately. As we quickly mounted our pursuit, I was amazed at how quickly I learned the difference between The Newbie's bawling and Ol' Major's. Maybe it was because he was still basically a post-pubescent pup, but it didn't have the deep base tones of Ol' Major's.

A few minutes into our flash-lit stumble fest across steep hills and through the thick underbrush, The Newbie's bawl began to crack like the voice of a teenage boy. We didn't think anything of it, but when the tone didn't return to normal, Jim got a worried look on his face. Oddly, the new tone sounded almost like crying. While Jim and Claude assessed the new sound, Ol' Major came trotting out of the dark woods, making a beeline to Claude. Now Jim reckoned something was seriously wrong. Very seriously wrong.

Long before I figured it out, Ol' Major, Claude, and Jim knew what the problem was. And before The Newbie returned to the clearing, none of us had any doubt—we were downwind. The Newbie had done the unthinkable for a coon hound—he had gotten SKUNKED.

It was a long, cold, silent hike back to the truck for Jim and The Newbie. It was an even longer drive out of Brownie Creek hollow for me. The Newbie had acquired solo accommodations in the bed of the pickup, thanks to his strong odor. Ol' Major was granted a senior's reprieve and joined us in the cab. So for the bumpy ride out of that hollow, Jim and I both polished our awkward gearshift avoidance techniques.

The next time I go pike fishing with Gerlof, I'll tell him my coon hunting story and explain the expression "getting skunked". I'm sure Dutch fishermen have their own word for not catching anything, but I'm pretty sure it won't be nearly as graphic as the smell of that coonhound!

FINAL LOSS OF INNOCENCE

I was never what you would call a "believer". I was skeptical and asked a lot of questions. Most of the great myths of life I debunked at a relatively early age. Even through the lens of childlike innocence, if you look just below the surface, you can see the fallacy of most myths. How could anyone believe some guy could travel around the world in one night dropping off presents to every child along the way? Hell, it's nearly impossible to fly to Chicago on time without your baggage getting lost. I knew my penchant for the truth would be tough on my folks, but we all had to get real. Santa bit the dust pretty early on, as did the Tooth Fairy.

I opted for the truth from an early age and without remorse. Dozens of other myths fell to the wayside: life is fair; the check is in the mail; eggplant is good... on and on. For almost 50 years, however, one myth stood the test of time. When I finally realized I had been misleading myself for all those years, it was more painful than being dumped by my first teenage heartthrob.

The building of this myth started in my early years. I've been camping since my youth in New Hampshire. During the summers, in the woods behind our houses, my buddies and I would build lean-tos out of branches and spruce boughs. After riding our bikes to the point of exhaustion, we'd fall asleep at our makeshift campsite. And there were always the more adventurous Boy Scout trips to Mission Ridge or Morey Pond on Mount Kearsarge, or to the cliff known as the Bulkhead and Huntoon pond on Ragged Mountain. On the Boy Scout trips we had some great times singing around the campfire at night, and racing back to the fire in the morning to get warm after leaving the friendly confines of our sleeping bags. As an adult, I continued the tradition of great camping and fishing trips. As each of my sons reached 8 or 10 years old, my Troop 281 companions were replaced by Ethan and Dan.

After the boys went off to college, I began a tradition of leading a group of friends and co-workers on our long Columbus Day weekend on an adventure to the Snowy Range of south-central Wyoming to fly fish. Boy Scout songs were replaced by ribald stories, card games, and several warming liquids other than hot chocolate. The morning rush to warm up by the campfire still took place, but it evolved to include a strong cup of coffee, a check of the weather, and a plan for the day's fishing. I always found several people to go along—I never camped by myself. Then one year, things changed.

The word had spread among my comrades over the years about the difficult weather conditions our group faced during this trip. Not all of these guys were as experienced campers as I was, after all. After a few years, everyone I tried to cajole into going with me had a good reason for turning down the adventure. One guy had to rearrange his sock drawer, and another said he was thinking about having a kidney transplant.

I'm a rather serious, some would say a fanatical, trout fisherman. In my zeal for big brown trout, I hardly noticed that the number of participants had dwindled. In retrospect, after each trip during which the weather turned miserable, someone dropped out and didn't sign up the next year. But when you're a fanatic, you don't notice little things like sub-freezing temperatures or major snowfalls. What matters is how good or bad the fishing turns out to be.

But fanatics die hard. Ten-plus years of trips had given me too many fine memories; I decided to go with just one trusty companion who would be there with me until the end: Sandydog, a 70-plus pound Terrier-Chow-something else mix. A dog with the personality of a California valley girl – for sure!

The workday was over, the truck was packed with all the gear and food my sole companion Sandydog and I needed for a long drive and a fishing adventure. With high spirits, we were on our way. For the first 25 miles, Sandydog was as excited as I was to start our adventure. She stood propped on the armrest on the door, her head out the window and her tail wagging a mile a minute. Eventually, the car ride made her sleepy, and her large frame splayed out across the seat and oozed ever more into my lap as she snoozed past every mile marker.

Eight hours on I-29 and I-80 and two gas stops later, we stopped for the night at a basic mom-and-pop generic motel in North Platte, NE. No need for dinner, we had been feeding on the official food of a road trip, Little Debbie Oatmeal Cream Pies, as we tooled along the way. Sandydog didn't see any difference between the bed in the motel room and the seat of the truck. She staked out her half from the middle and went to sleep. Once I managed to claim my section of the bed, sleep and dreams of big brown trout came quickly.

The sound of my 5:30 wake-up call spooked a big brown that was rising to my perfect size 22 Olive Bodied Adams fly and jarred me awake. Within minutes, we were back on the road. Breakfast was a cup of motel lobby coffee and the homemade oatmeal cookies my beloved had secretly packed away for me. Sandydog's enthusiasm for the trip was renewed at least until the last cookie was gone. We headed toward the western horizon on the familiar straightaway. In just six more hours we would be fishing!

Just after 11 am, we finally crossed the Libby Flats, just west of Centennial, WY, and descended into the Hog Park Creek valley. Barring any travel issues, we would reach Hog Park Creek and be fishing before 1 pm. Hog Park Creek is one of those lazy mountain meadow streams. With all its meandering, twists, and turns I bet it travels five miles to cover the valley's three-mile length. About halfway through the valley, Hog Park Creek takes a fancy to the hills on the south side of the valley. There, it bubbles past a stubborn rock outcropping that didn't yield to the force of the creek as easily as the others, producing a stunning ponderosa pine-covered point. From this commanding lookout, the only high spot in the valley, you can peer into the creek and watch trout rising in the smooth water of a beaver pond. With trees for shelter and firewood, trout rising at the doorstep, and the warm morning sun as an alarm clock, what more could a guy ask for? No wonder so many, myself, my sons, and my former fair-weather fishing companions included, had chosen this for a campsite over the years. Even the local prehistoric Indians had favored the spot for its beauty and the beautiful quartz stones they could fashion into tools and gifts of jewelry. But this year I could not camp there myself. My history-filled campsite had been declared an archeological site, and was now off limits!

Just down the valley a few hundred yards to the east was the "primo" fishing area, the mouth of a secondary draw where the South Fork entered the main Hog Park Creek. While not nearly as nice as my pine-covered point, a new camping spot near the junction of the valleys had several redeeming qualities. It was flat and had a nice rock to include in my fire ring, and it was only steps from great fishing. Despite the lack of trees for cover and wood, and no early sun or stellar view, it would do. After all, I could damn near cast to a trout from my sleeping bag.

By 1:00 I had pitched my tent and rigged my rod and was headed the short way to the stream. A short ten minutes later I was fishing. And what a beautiful day for fishing it was. No wind, clear skies, 60-degree temperatures, and a caddis fly hatch! Within five minutes, Sandydog, exhausted from sleeping all the way, found herself a spot in the sun and left the fishing to me. As the sun finally warmed the water, and me, the caddis hatch intensified, and the brown trout started to rise. I forgot all about being alone. My only thought was of catching yet another one of those big browns that had been sipping in my fly. But, after a few hours entirely engrossed in my passion, the sun sank behind the mountains lining the valley on both sides of the stream and the temperatures cooled rapidly. I had forgotten that all I had done was pitch my tent.

Coming to my senses at about 4:00, I remembered I had to lay in a supply of firewood and attend to other camp chores. At the new campsite in the middle of a mountain meadow, trees, and thus firewood, were in short supply. By the time the sun had set, I managed to scavenge a barely adequate supply of firewood. A drawback of this solitary style of camping became obvious: many hands make light work. It is funny how it takes the same amount of wood to build a fire for four or five people as it does for just one.

By the light of my lantern, I got a fire going and prepared Sandydog and myself a meal of Chicken Marsala and pasta with pesto sauce. It was getting cold, even hunched over the cooking fire. I put on a flannel shirt, followed shortly by my coveralls. It was the start of a clear, cold autumn night in the mountains. Stoking the fire to keep warm and hasten our dinner preparation, I realized a long evening lay ahead. Sandydog is all you could ask for in a pet, but she's wasn't a great conversationalist. There would be no camaraderie and storytelling tonight. There would be no good-natured teasing of other companions about the fish they didn't catch, or telling ghost stories to Ethan and Dan.

I thought, "Get dinner over with and go to bed early." It sounded like a great idea. Ben Franklin's words floated into my head, "Early to bed, early to rise." I wondered if he ever fished.

It doesn't take much Marsala to make Chicken Marsala, even if you are making it for five hungry fishermen and a lazy dog. When there is only one fisherman and a lazy dog it takes even less. I had an entire bottle, and only needed about half a cup. With dinner finished and the dishes done, I huddled near the fire to try and keep warm. Sandydog looked at me sympathetically. She couldn't tell stories, jokes, or play cards, and she knew it. All I could do is stare at the coals of the dwindling fire, and recall past trips; both the good ones and the ones that tested me because of tough conditions.

Again I thought, "Go to bed." But I was afraid if I did, I'd wake up too early in the morning. What to do now? Cold and no human companionship. I could think of only two things: feel sorry for myself, and drink the bottle of Marsala. I did both, and as the night grew colder, the wood pile grew smaller and the bottle of Marsala lighter. It might have been freezing, but Sandydog didn't seem to mind. She found a spot near the fire, rolled herself into a little ball, and tucked here tail around her nose. She was even snoring a bit. For me, though, enough was enough, both of the Marsala and the bone-chilling cold. So what if it was just a few seconds past eight? I'd worry about waking up early when it happened. Without delay, I was off to my tent and the warm confines of my down sleeping bag: no "passing Go", no "collecting $200", and no assembling tinder, kindling, and the other necessities of a morning campfire. Just get into the warmth of my sleeping bag and get some sleep.

The warmth of my sleeping bag and the bottle of Marsala made for a cozy night. There was no need to count rising trout. I fell fast asleep, likely making my own sleep noises, in no time.

At about 1am, I started to get a squirmy feeling and heard a scratching sound. The campfire had burned down, and Sandydog wanted in the warm tent with me. At the same time, I realized the Marsala I had so greedily consumed by the fire wanted out! Reluctantly climbing out of my toasty down bag, I slipped on my camp shoes and unzipped the tent flap for Sandydog and the Marsala. Out I stepped into a world of cold and frost. Mist from the Hog Park Creek had turned into a thick, snow-like coating of frost over everything, including Sandydog. Moving only as far away from the tent as necessary, I let the Marsala out as quickly as possible and hurried back into the tent. I barely paused long enough to glance at the campfire. All I saw were a few small glowing embers, and all I heard was the quiet of a cold mountain night and the gurgle of the nearby creek. No blazing fire to warm anyone, and none of the laughter that follows tales told beginning with the phrase, "Did I ever tell you about the three guys . . ." The effigies of camping companions around decades of blazing fires seemed both present and conspicuously absent. Cold and alone, I hurried back to my sleeping bag and Sandydog in my tent.

I was awake long before morning's first light. The long drive and day of hard fishing hadn't compensated for having climbed into my sleeping bag at 8 pm. I tried to roll over and go back to sleep for just a little longer, but my Marsala-induced headache wouldn't let me. I didn't want to look at my watch and see how early it was. I just wanted to stay in my toasty sleeping bag, have my head stop pounding, and hear the sounds of a bustling camp, sounds that would mean someone else had arisen early to stoke the fire and start the coffee brewing. Mustering all my resolve, I pulled on my jeans and sweater and unzipped the tent flap. Not a sound could be heard from the campfire, and no camp sounds translated into no coffee brewing. God, did my throbbing head need coffee.

Why wasn't there a fire in the campfire spot? Sure, I hadn't neatly laid out tinder and kindling, but both were near the campfire, and there were still enough logs left to build a fire to perk the coffee. But, the fire hadn't been started. The embers from the night's fire were gone, and the ashes were as cold as the pre-dawn air.

There is always a fire going when I climb out of my tent in the morning. How could this steadfast law fade into myth so unceremoniously? The cruel reality hit me at that moment right in my throbbing head. There was always a fire going in the morning because someone else had climbed out of their tent well before I got up--and they started the fire and got the coffee brewing.

In that cold awakening I faced the reality and hard fact of life. No one, like a wisp of smoke riding on the wind, was silently flitting from camp to camp in the early morning hours carefully putting kindling in the campfire spot, laying kindling over the tinder, topping it with a tent of small logs, and artfully striking a single match. No Santa Claus, no Easter Bunny, and even worse, no Campfire Fairy! Alone and with a shiver, more from loneliness than the cold, I surrendered my last great mythical belief. My last vestige of innocence disappeared.

The difference between camping with my friends and family and camping alone opened a window into a world apart. A consummate naturalist, I always thought the beauty of the mountains and water was the defining joy of the outdoor experience. But that morning I learned something that has resonated deeply within me ever since…. It's about the people you share it with.

NORM, NOT NORMAN

Who we are is reflected in our name; or, maybe it's the other way around and our name reflects who we are. Either way, it defines us. In this phase of my life, I'm Norm. Norm is a shorter version, a diminutive, for Norman. Norm is always within Norman. Another name, like perhaps Ralph, that can't be shortened, might have made my life a little simpler. Maybe I wouldn't have had to come out of the personality closet—but oh, what a long, strange, and at times, wonderful and invigorating adventure it's been!

I was born, as my birth certificate says, Norman Harley Crisp on August 6, 1947. On October 2, 1955, I was confirmed as Norman Crisp in the Episcopal Church, or as we called it, the Angry Anglican Catholic Church. My Grandfather confirmed it in his inscription in the Book of Common Prayer he gave me on that auspicious day. He also wrote, "May God be good to you." My name was officially Norman.

I think many of us go through our lives never exactly knowing who we are. Sure, we see ourselves in the morning when we brush our teeth, or when we take a selfie. We have idealized portraits of ourselves, but how accurate are the mirrors of whom and what we are? Does seeing the "touch of gray" make us older or distinguished? I suspect that our introspections are somewhat different than what others perceive, and that the perception varies among the people with whom we come in contact. So how do we define ourselves? Do we fabricate a persona of who we want to be because we don't like who we are? Do we project a persona based on how we perceive others as seeing us? Or maybe, if we are lucky, wise, and the time is right, we discover the persona resting in the name to which we want to respond. I'm slowly coming to know who a Norm is, in large part, by coming to know I can't be a Norman.

I didn't have too difficult a time being Norman for the first part of my life. You hear your family tell you, "Norman, dinner is ready," or ask, "Norman, did you do your homework?" and that is who you are. Norman was my parents' child, the Marine Corps' property, and my wife's husband. As long as I was those things, Norman appeared to reflect the person I saw every morning as I brushed my teeth. When I wasn't any of those anymore, the persona looking out through the fog that formed on the mirror from the steamy moisture of my morning shower began to morph into someone new.

About 25 years ago I started my transformation from Norman Norm. I didn't even know it was happening. It would be easy to use the analogy of a snake shedding its skin. The snake grows, its skin becomes too tight, and it wiggles out of the old skin. But it still has the same form, and is the same species of snake, just a little bigger. As an aquatic biologist and avid fly fisherman, I think a better analogy lies with the caddis fly *Rycophilia sp.*, also known as the green rock worm. Like *Rycophilia sp.*, I was going through a complete metamorphosis. The larva and pupa stages of the green rock worm look different and act very differently from each other. Finally, the pupa emerges as something best imitated by a size 14 olive Elk Hair Caddis. It's still a *Rycophilia sp.*, but nothing like a green rock worm or the pupa. Once you emerge, you can't go back.

Norman was a nice guy. You would like him if you met him. Bob Segar must have met him because Norman was the kind of person that Bob described in "Beautiful Loser": "He's always willing to be second best, a perfect lodger, a perfect guest . . . He'll always ask, he'll always say please." Norman was not a cheerless person, he just wasn't terribly joyous. Norman worked hard at having order and precision in his everyday life; it wasn't until later that he learned his highest need was actually for change, and his lowest need was for order. He was loving and caring, but you couldn't find it in his voice. Every once in awhile, Norman showed a hint of fun, excitement, and spontaneity, but it was just a hint and it was easily scared off. Like a box turtle, he had his carapace to pull his head back into if things got too risky. But even when safely tucked in his shell, he knew something didn't feel quite right. In T. H. White's book *The Once and Future King*, Merlin turns Wart, as he called young Arthur, into a goose. Once returned to his human state, Arthur tells Merlin that he keeps having this feeling that there is something he is supposed to do. Once he started his migration north, the feeling started to go away. Once Norman started to migrate toward Norm, a lot of old feelings fell away, and new feelings began to fill the void.

Beyond nature and nurture, I began to expand and deepen. Norm is the result. It didn't happen suddenly like flipping a switch, but once the process started, it followed Newton's first law of motion—I became an object in transit with direction and speed, unencumbered. And here I have arrived, still traveling. I am Norm, not Norman. These days, it is an easy pleasure to be Norm; I know how to be Norm. It is as simple as breathing, and breathing is easy. Once in a while I may get a chest cold, but I keep on breathing, and before very long I'm back to Normal.

Who have I become and how have I changed? Norm is the guy who wants to know what is around the next bend in the road. Norm loves everything about food: buying it, cooking it, and enjoying it with friends and loved ones. He enjoys the ballet and symphony without always having to remember who was the choreographer or composer, or make small talk about the performance. Norm is content to piddle away time lying on his back looking at the sky and finding "cloud bears".

Norm also contemplates what some stream water quality data means as he takes his morning shower, or what he can do differently to be the best possible dad in each stage of his sons' lives. Oxymoron notwithstanding, Norm is a responsible free spirit, a kind of a reverse Tootsie Pop. He's soft and spirited on the outside, and firm and contemplative on the inside. And like a Tootsie Pop, without both components, it is just not the right kind of candy.

As Norm, I've reached places I never really knew existed. Once, getting extremely sick after a trip to Morocco, I thought I was going to die. I realized I had a fair share of sadness about some things that happened along the way, but I found joy in knowing I had no regrets of things I did, or could have done, but didn't do.

Now, and through the transition and unfolding that has shortened Norman to Norm, Grandpa Crisp's prayer and wish has come true. God has been good to me! And I'm not even a believer.

IT'S GENETIC

"If an urge to explore rises in us innately, perhaps its foundation lies within our genome. In fact, there is a mutation that pops up frequently in such discussions: a variant of a gene called DRD4, which helps control dopamine, a chemical brain messenger important in learning and reward. Researchers have repeatedly tied the variant, known as DRD4-7R and carried by roughly 20 percent of all humans, to curiosity and restlessness. Dozens of human studies have found that 7R makes people more likely to take risks; explore new places, ideas, foods, relationships, drugs, or sexual opportunities; and generally embrace movement, change, and adventure. Animal studies, simulating 7R's actions, suggest it increases their taste for both movement and novelty. (Not incidentally, it is also closely associated with ADHD.)" National Geographic, January 2013

I found out I am a DRD4-7R variant accidently when I was in my early 40s. In retrospect, I recognize some of the symptoms. A personality test I took years ago indicated my highest need was for change, and my lowest need was for order. For as far back as I can remember, I would try any type of new food, embrace new ideas, act spontaneously, and take risks in my life choices. (Risk-taking and thrill-seeking are two different things—Risk? Yes. Thrill? No.) And, I was diagnosed with adult ADD.

The genetic variant most certainly came from my dad's side of the family. My great-grandfather, Oliver Crisp, immigrated to the US from Wales sometime in the early 1870's. My Dad likely received the gene. Sometime in the early 1930's, Dad and two friends drove to California. That trip must have been an adventure. Dad never volunteered to talk much about his adventures along the way. But when the subject came up, his eyes got that special look you only get when you remember something significant, like kissing your wife right after you have been pronounced man and wife by the minister, or holding your child for the first time. Dad never took another trip because the business of life got in the way, but he passed the DRD4-7R variant to me (along with the sex-linked holandric gene for hairy ears), and I passed it along to one of my sons. Dan takes pride in having as many, or more exotic visas and stamps in his passport than I have in mine.

Of course, I always had 7R variant characteristics, but they remained hidden in my ho-hum daily work world and suburban life, much as the variant was suppressed within my dad. Then in 1983, at age 36, my life changed, I divorced. I started on a new, irreversible course; the change was palpable. I did things I had wanted to do before but never felt free to undertake. I pursued an interest in photography, rekindled my love of fly fishing, and cooked foods I had always wanted to taste. With each passing year, an urge deep within me increased in intensity. I began to understand what a goose must feel as the days grow longer and spring approaches, when deep within its soul it knows it is time to fly north. In 1989 I started my migration. That year, I took my first adventure. I went to New Zealand by myself. And things were never quite the same again.

I had a perfect job, at least for me. In the warm weather, I traveled throughout Kansas, Nebraska, Iowa, and Missouri doing water quality studies: sampling fish populations, collecting water samples, or doing special projects. Generally the fall and winter were spent writing project reports. In the summer of 1989, my co-worker Katie and I spent most of the sampling season on the road doing these studies. We shared a lot of travel time, driving through the Sand Hills of Nebraska and the Missouri Ozarks to our study locations, and we spent most of it talking. After college, Katie had spent the better part of a year traveling around Europe, and inevitably, she relived her trip during our long drives.

Ultimately, I started thinking maybe I should do something like that. I didn't have stories of adventures to share with her then, but I did have an unspoken yearning to see New Zealand and to fly fish for its fabled trout. For fly fishermen, New Zealand, along with Patagonia, is one of the sacred pilgrimage locations.

One day, out of the blue, I said it: "If we receive performance awards for all this fun we've had collecting water samples and seining fish this summer, and the payout is enough for a plane ticket to Auckland, I'm going."

Katie replied, "Okay, deal. If we get good performance rewards, I'm going to get a new guitar."

"Deal," I replied.

That September, we did get nice performance awards. I hadn't forgotten my vow that I had made with Katie that day, but I hoped she had. The thought of going to New Zealand solo was exhilarating, but as I thought things through, it was scary, too.

Katie hadn't forgotten. On our first trip together after our performance reviews, she brought her new guitar and a calendar. She threatened to tell everyone I chickened out and even to play the guitar out of tune until I wrote my trip dates in her calendar. She wasn't fooling. Still, I was reluctant, very reluctant. Going off on this first adventure was terrifying to me. Sure, talk is cheap, but actually doing what you talk about is something very different. Yes, it was something I had to do, just as the geese have to migrate north. But to the geese, it is an innate trip, completed without thinking or any second thoughts. For me, the discomfort of thinking about this sudden change in lifestyle was far greater. The internal flutters throughout my entire body shook me up every time I thought about what an experience it would be. But Katie was ruthless. She immediately began to purposely play out of tune, and kept this act up the entire trip. Worse, during our first post-fieldwork staff meeting, she dropped the bomb on me. I gave what I thought was a succinct summary of our accomplishments, focusing on our success in refining our "mixing zone" delimitation procedures. I was feeling rather smug about my "how I spent my summer vacation" report and ready to humbly accept my coworkers' accolades. But before I could bask in our glory, the section chief, with a glimmer in his eyes, asked Katie if she had anything to add to my report. I thought, "Oh shit, here it comes."

Katie added a few comments about calculations of dye injection rates. My chest was still tight but loosening by the second. Then it came. As we all reassembled our notes and prepared to head back to our cubicles, she interjected, "Don't schedule Norm for any field work in November. He told me he was going to take his dream trip to New Zealand then."

And just like that I went from being just another cubical neighbor to being the hope and hero of all the other cubies.

I had to get past the hump of hesitation to rid myself of the cloud of shame hanging over my head, as well as to escape Katie's agonizing off-key renditions of "Me and Bobby McGee". I made the reservations, and my life was changed forever. Damn you, Katie-- and thank you.

NEW ZEALAND – JULIA THE WITCH, THE FIRST ADVENTURE

As soon as Katie spilled the beans at the staff meeting, I knew the die had been cast. Now it was time to start planning for this once-in-a-lifetime trip. I decided I'd go during the New Zealand spring: late October into early November. Perhaps I should have waited until New Zealand summer for the trip, but that would have meant significantly higher airfare, plus a greater chance I would chicken out. I couldn't bear the thought of enduring the ribbing Katie had promised if I didn't follow through.

My first order of business was to find a place to stay. Since I'm a frugal soul, there was no way I was going to spring for a fishing lodge that would cost close to $1,000 per week (the going rate, in US dollars, in 2016). Hell, I was budgeting $50 a day. Besides, if I was going halfway around the world, I wanted to see more than the immediate confines of the lodge and its environs. New Zealand is two islands, about 950 miles from north to south, and about one quarter again the size of my home state, Kansas. If I was going to muster the courage to go there, I was determined to see as much of the country as I could. Little did I know I was setting the pattern for all my future travels. Fishing would be my excuse for going someplace new and exciting. I would fish on trips, not just go on a fishing trip.

The second decision I had to make was, with such a big country to fish, where would I start? Being resourceful, I did the obvious. I started going to bookstores and clandestinely reading fly fishing magazines and scouring the classifieds. No one appeared to notice me doing my research, maybe because I avoided the store's high traffic areas. I kept my sleuthing quick, making cryptic notes about lodges and locations. This was in pre-Google days; you had to work for information. Most of the lodges I found had a US phone number or address. I'd copy the information, then call or send a letter asking for information about their programs. I got very little in return for my efforts: just their rates, general location, and where I would fly into to be picked up. I was able to deduce that the Nelson area on the South Island and Lakes Taupo and Rotorua on the North Island were fishing locations, but I would need a lot more information than that for a frugal fly fishing trip.

One day as I sat in my cubicle eating my lunch of yogurt and fruit and mentally planning my trip, I had an epiphany: I would contact Trout Unlimited, the premiere cold water conservation organization in the US, with state and local chapters throughout the country. Surely a country with arguably the world's best trout fishing would have TU chapters. Unable to find contact information on my own, I called the national office in Washington, D.C., for help.

My call was answered by a temporary employee who was manning the phones while the receptionist ate lunch. She had no idea how to answer my question, but transferred me to the executive director. Sure enough, he had just the information I needed! He informed me New Zealand has their own organization called "Trout Unlimited New Zealand". It isn't officially affiliated with the US TU, but they share common goals and objectives. The two are basically siblings from different parents.

The executive director had nothing on his plate just then except for his lunch, and I had nothing going on either. We ended up having a wide-ranging conversation about fly fishing, trout, and life. Our conversation resulted in the motherlode of information I was seeking: the name, phone number, and address of TU New Zealand's executive director! Once our conversation ended, I immediately wrote the New Zealand TU executive director, explaining how I had received his name and address, telling him of my upcoming trip, and asking for his help and guidance. For additional impact I printed the letter on EPA Region VII letterhead. That, along

35

with my reference to the executive director of the US TU, must have given the impression that I had some "juice". Not long after I mailed my letter I received a late night phone call. It was New Zealand TU's executive director telling me he was looking forward to meeting me. He promised to meet my incoming flight if I sent him the information.

The Kiwi people are friendly and accommodating. True to his word, the executive director of New Zealand TU and his fishing mate met me at the airport. He took me to his house and showed me to my room for the night. I felt like I was visiting fly fishing royalty. No doubt the "blessings" of the US head of TU and my official stationary facilitated my welcome. After a nap and a traditional Kiwi dinner of roasted lamb, root vegetables, GOOD wine, and of course, sliced kiwi fruit for dessert, the table was cleared and a map of New Zealand produced. Over more wine, the director suggested a fly fishing tour route of both islands. At each fishing location he had penned in the name and phone number of a nearby New Zealand TU member. There is nothing quite like local knowledge. Armed with this insider trading info, I set off on my first great adventure.

The first couple of days I fished alone, had a beer or two at a local pub, and slept in hostels. Then, my route took me to Taihape and the Rangitikei River, the first location with any contact information. On the map beside this location was written a first initial, a last name, and a phone number. After my routine evening pint of Lion Red and a sausage pie, I called to introduce myself and explain how I had received the number.

A woman's voice answered my call. Female fly fishers were rare back in those days so I told her who I was looking for. Her reply was, "It is me." My explanation apparently satisfied her. After a few perfunctory questions I was given directions to her house and told if I hadn't eaten yet, don't. A venison roast was about ready to come out of the oven. Twenty minutes later I was poised to knock on her door.

A petite woman a few years younger than me answered. She wore jeans and a short-sleeved shirt. She had long, flowing black hair and crescent moon earrings dangled from her earlobes. Tattoos on her forearms also caught my eye: a "Jack Scott" salmon fly, and a pentagram with the moon in the center. (I would later learn she had others; you needed a mirror to see them all at one time.) As she walked me through the front room toward the kitchen and a platter of roasted venison, parsnips, and beet root, I noticed what appeared to be an altar.

During my first helping of dinner, conversation revolved around getting to know each other a bit. I learned she was a sheep shearer, single, who subsistence hunted, and didn't fly fish as much as she spent her time tying classic salmon flies. That explained the "Jack Scott" on her left forearm. During my second helping of venison she casually said, "I'm a witch." That explained the pentagram on her right forearm and what appeared to be an altar in the front room.

As I was finishing my after-dinner herbal tea, she excused herself to prepare the front room for evening devotionals. After lighting incense and candles she went to change. I wasn't sure what to expect. Would she come back with a pointed hat, black cat, and broom? No, she just changed into a long black robe. Devotionals were a simple ceremony of waving some flowers through the incense and chanting some softly spoken words as she faced, in turn, toward each of the compass points. Being a-religious, I watched her ceremony with detached interest, much as I would have watched an Episcopal confirmation. Once the devotionals were over, we went back to the kitchen for some apple pie and too much wine. The night ended with a tour of the rest of the house—and her tattoos. All good things must end, and after two days of consorting it was time for me to move on. But, before I departed she led me to her alter and started an incantation to conjure up the trout of New Zealand. As she waved incense around me she told them it would be okay for me to catch them since I would honor them and release them. And with that, I went on my way.

The fishing got much better after I left Taihape. I'm not sure how much the incantation helped. It may have been coincidence or maybe my future locations were just better places to fish, but then again, maybe not. Regardless, it was a memorable first adventure. I can still picture vividly my experiences climbing on a glacier and camping under the Southern Cross. But nothing else on my trip would compare to evening devotionals and consorting with a witch. A wanderlust spell had been cast over me. On the flight home I realized I had not spent nearly what I had budgeted for this initial adventure. I knew then, more trips were in my future.

PAKISTAN CUSTOMS

A man will go to great lengths to impress the object of his affection. For me, that object was a woman I met at a second grade back-to-school night. I was sitting in my son's chair with my knees bumping the bottom of the desk, when I caught the eye of an attractive woman on the other side of the room. It was one of those unexpected encounters where you both keep casting quick glances at each other. When the "class" was over, I lingered in the hall outside the room waiting for her to come out. I don't remember what I said, but it was probably dumb. It must have had some redeeming quality, however, because we bantered for a while. We finally parted ways, her heels clicking on the tile floor and her ass swishing as she strode down the hall. We met again on a second grade field trip. That cemented our attraction to each other.

We started to meet on Sunday nights. After the kids had gone to their other parents' houses we would have a simple but sensual dinner with rose-scented candles, our favorite aroma for a candlelight dinner as well as massage oil. She enjoyed a good massage about as much as I enjoyed playing masseuse. Adding essential oils to the mix added to our enjoyment. Now, you can purchase essential oils at any health store or couples shop. But I was looking for something a little more exotic; something with a little je ne sais quoi. This was going to take just the right connection.

I decided it was time to pay my nephew Jeff a visit. It was 1997, and I hadn't seen him since Memorial Day 1990 when we both happened to be back home in New Hampshire. At that time, Jeff had just returned from the Peace Corps in Nepal and I was just starting on my life as a part-time vagabond. We had been rather close in years past, but when he moved to Kathmandu we lost track of each other. Now seemed to be just the right time to rekindle our relationship. Jeff was involved with helping a Nepalese community on the Tibetan border high in the Himalayan Mountains extract and market essential oils from flowers. Getting exotic essential oils from their source was as good a reason as any to fly halfway around the world, and a visit with Jeff would sweeten the deal.

I'm a frugal traveler, and I can almost always find the cheapest travel deals. That doesn't necessarily mean the best accommodations or the most direct flights. I always traveled with my sleeping bag, just in case. Most of the time I didn't need it because I could find a hostel or one star hotel. But, it was nice to have when I had to settle for a barn or under a bridge. There were not, and still are not, any direct flights to Nepal from the United States. I booked a flight on Pakistan International Airlines (PIA), which at that time only had one flight a week to or from the U.S. I would have to change planes three times along the way: in Amsterdam, Lahore, and finally, Karachi. Pakistan International Airlines is not a major player in the U.S. to Middle East and Asia air routes. This becomes apparent upon arrival at the airport. With such limited service, you can't afford counter space in the international departure area. Instead, you have something just a step or two above a couple of folding tables in a metal building, and I guarantee it will be a long hike or short taxi ride from the main terminal. But, PIA had the lowest fare to Kathmandu, and that's all I needed. The plane was very old, the carpets were threadbare, some of the overhead bins wouldn't open, and my seat was stuck in the upright position. I expected it to smell like most flights—the air stale, recycled. But the first thing that struck me on this flight was the smell of a curry-like spice. The crew was also atypical of most flight crews. Instead of being made up of mostly young, attractive women, they were all men, and all had beards. I knew very little of Islam and the Muslim world. It never occurred to me that a woman would not be allowed to be a flight attendant, nor did I anticipate the absence of alcohol on the flight. As we pulled away from the gate, the flight steward pointed in the general direction of Mecca and said a

Koranic prayer for our safe journey. I was fine with that. On an old plane any prayer is potentially a good prayer. He finished, then announced to the passengers, "Drug trafficking is a capital offense, and you are responsible for anything in your possession." It was almost an afterthought. I was on my way to Kathmandu, considered by some the home of some of the best weed in the world. Why would I bring my own?

This leg of the journey was relatively uneventful. I passed the time by thumbing through my now well-worn copy of *Nepal on $20.00 A Day*, and wishing I could have a glass of wine. We made stops and plane changes in Amsterdam and Lahore. I was bone tired when we finally reached Karachi. I gathered my bag and joined my fellow passengers at the "In Transit" desk. They kept us waiting another 15 minutes after we had all arrived, likely just for good measure, and then herded us out of the airport and onto a waiting bus. The building had been hot and muggy, but the night air brought no relief. Instead, the heat and humidity increased its assault, instantly drenching my body in sweat.

I turned my attention to the multitude of families apparently living on small, maybe 10 X 10 foot, woven mats, tucked between the curb and the airport building. Somehow in that small space an entire family found room to eat, sleep, and live. The lack of privacy, food, and water made the situation all the more depressing. Though my small room in the transit hotel lacked grand amenities, I at least had an overhead fan and a place to take care of basic bodily functions. It made me realize that being of American birth was akin to winning the lottery. I couldn't imagine how they endured the heat. How they managed to live that way was beyond my comprehension. Welcome to the "developing world".

A few short hours of sleep later I dined on a breakfast of some unknown but delicious pale yellow fruit that I later learned was jackfruit. It was accompanied by rice gruel and a fantastic tea. Then, it was off to the airport for the final leg of my flight. Pakistani customs gave my bags and passport a quick once-over, and I was on my way to re-check my bag, receive my seat assignment, and wait for the boarding call for Kathmandu. The airport "city" had been a rude awakening to me. Would Kathmandu be like that? What about the countryside? A couple more hours and I'd know.

After Karachi, Kathmandu and the Nepalese countryside was an oasis. With my bag claimed and a cool new visa in my passport, I walked out of the airport to find Jeff waiting for me. The heat and humidity were nothing compared to Karachi; in fact, I found it rather tolerable. The streets were crowded, but not teeming with throngs of humanity, nor did I see families living on the airport grounds. The worst things I had to endure were the acrid smell of diesel exhaust and having to step carefully to avoid the trail left by the sacred cows.

In Nepal, miles and kilometers don't hold much meaning. Instead, distances are measured in days, as in how many days it takes to get to your destination. Most of the times, walking days are the standard measure. On occasion, it refers to flying days. That doesn't necessarily mean days in the air. It includes how many days you have to wait for one of the 19 seats on the Royal Air Nepal De Havilland Twin Otters plane. It might also include the time you have to wait for a second seat should you be bumped in favor of a sack of rice or other important commodity. That would add up to another seven days.

Jeff and I were headed for Simikot in the Humla District, the mountainous region of northwest Nepal bordering Tibet. That's where we would find the essential oil extraction facility and my anticipated mother lode of essential oils. They wouldn't be rose oil. The oils would be extracted from the hand-picked blossoms of rhododendrons and other flowers growing in the Himalayas. The trek to retrieve the oil would make the aroma all the more amazing.

The first leg of the trip to Simikot was simple. We hired a car and driver to take us to Nepalgunj, a border town between Nepal and India. From Nepalgunj we had two options: we could walk for several weeks, or we could take the once-a-week flight. We opted for the flight. Somehow, we managed to secure the last two of the 19 seats available. I think we paid a "special surcharge"; the details are a bit fuzzy, if you know what I mean. As we prepared to board, Jeff, who had seat 19, was stopped in his tracks. He'd been replaced by a 50 kilogram bag of rice. Before Jeff headed back to the terminal, he gave me this advice: "When you arrive, just say "Jeff" to any official-looking person. See you when you get back to Kathmandu." With that he was gone, and I settled in next to my new rice seatmate.

The flight to Simikot was breathtaking. The route follows a river valley with the Himalayas towering on either side. Simikot is at about 9,000 feet altitude, perched on the side of the mountains. The runway is only about 1,800 feet long, making both takeoff and landing an adventure in and of itself! In fact, the landing was even more awe-inspiring than the flight. With such a short runway you land going uphill to slow the plane down before it reaches the end of the runway.

As promised, "Jeff" was the magic word in Simikot. I picked an official-looking man and tried it out. He wasn't much help; apparently the only English he knew was "Jeff". But he quickly passed me off to an even more official-looking gentleman who spoke enough English to get me to a room in a small guest house in the midst of renovation. He also introduced me to the family in a house across the street and notified them I'd be eating with them. I spent a glorious week hanging out on the side of a mountain, watching the locals. The family who fed me consisted of three generations of women: a grandmother, daughter, and granddaughter. I speak only English (though some might consider my New Hampshire colloquialisms a bit foreign), and they only spoke Nepalese. In my experience, if you don't speak the local language, people think of you as a baby and feel free to say anything in front of you. But, I do speak universal "body language". Sitting quietly in the corner, watching the women share the cooking space and duties, I was able to translate their actions even though I couldn't understand their words. The mother was trying to be respectful to the grandmother while simultaneously asserting her dominance. The daughter, a typical teenager, poked her mother at every opportunity while looking to her grandmother for approval. No doubt if I understood Nepalese, they would have also hidden their body language.

The time came to make the trek from the village to the essential oil extraction facility. It seemed like it was a thousand miles down the gorge from Simikot. It was simple to say the least. It was an old boiler that produces steam that was passed through a vat of flowers. The steam was captured, and volatile oils condensed as layers on top of the water. It was a crude process with a very low capture efficiency rate. A lot of the volatiles escaped into the air. Rhododendrons were in bloom and being processed. The area around the boiler, in spite of the smoke, smelled what I imagined the massage oil would smell like.

While there, I also got to experience some of the local culture. Simikot is the nearest village to the Tibetan border, something I learned when I strayed too far from the village and was intercepted by a Nepalese Army troop.

I also learned that Buddhist monks and nuns escaping from Tibet to seek the Dalai Lama wait in Simikot, hoping to get a seat on the weekly plane. Apparently the schedule is widely known; a group arrived the day before the flight I planned to take. As they waited, they huddled around a small fire, drank milk tea, and played some sort of card game. Being curious by nature, I watched the game with interest as I whittled on a stick with a Swiss Army knife. Thinking I was starting to understand the game, I moved in closer to watch. While there was gambling involved, it seemed they were taking turns winning and passing their bets back and forth between themselves. Intrigued, I invited myself to join with a few simple gestures: pointing to myself, pretending I was dealing cards, and shaking my head yes. It worked. Two monks made room for me to sit down. The first couple of hands were no bet hands. I "won" one of them. Then the bets were on. I won again: a tea cup and a prayer flag. What I really wanted was a set of prayer beads. I had what I thought was a winning hand, so I went all-in with the Swiss Army knife. This is what they had been waiting for. Talking quickly among themselves, smiling and chuckling, they motioned that there was a slight rule variation of which I wasn't aware. My "winning hand" was actually a loser. They took the

Swiss Army knife (which actually belonged to Jeff), and I did finally win prayer beads. Maybe it was karma, maybe not.

The time finally came to say goodbye. Loaded with my booty from a week on the Tibetan border, Buddhist prayer beads, two bottles of Chinese apple brandy, and four small vials of light golden-colored essential oil from different Himalayan mountain flowers, I walked toward the plane terminal. My newfound gambling associates didn't make the flight, but they did come to see me off. Bundled up in heavy robes to keep the mountain chill at bay, they greeted me with palms together under their chins, bowed, and softly whispered, "Namaste." As I returned their gestures, the youngest monk reached into his wrap and, with a broad smile on his face, extended his hand with Jeff's Swiss Army knife on his palm. I shook my head "no," but they just giggled and smiled. The young monk remained with this hand out. As I fumbled in my pocket for the prayer beads I'd won, they all smiled and shook their heads "no." We may have been linguistically challenged, but we were all people.

My backpack and I made the exit for Kathmandu without any problems. I landed in Karachi, Pakistan. The Karachi terminal was less crowded than I envisioned it might be, so I made a quick and painless trip to my gate. I approached the x-ray machine and metal detector. This was before TSA restrictions forbade large bottles of liquids and carry-on pocket knives. Two bottles of Chinese apple brandy, four vials of essential oils, and a bottle of Nepalese catsup were just fine in your carry on. The issue was not limiting your liquids to three ounces, but how you packed your booty in your backpack so it didn't break. My essential oils were in one dram bottles, each container about the size of my little finger. Wanting to keep the vials together and offer some protection in case they leaked, I stashed them in my frugal traveler's Dopp kit, a Ziploc bag. I'd made room for my treasures by purchasing thrift store clothing before the trip. I used the best of it to wrap my prizes and tossed the rest. The catsup and apple brandy bottles were secured with dirty underwear, socks, and pants, and packed safely way.

I expected to run my bag through the scanner, sling it on my back, and leisurely stroll to gate nine. The scanner was manned by two uniformed Pakistanis. The scanner operator, obviously the junior member of the team, wore a rumpled uniform that looked like an unmade bed. The senior member of the team was his opposite: neatly dressed in a crisp shirt and trousers tucked into shined boots, his uniform decorated with ribbons and braids and a jaunty beret, his mustache carefully trimmed. All he lacked was a riding crop. He would definitely have been a finalist in a Saddam Hussein lookalike contest, and he knew it. He stood ramrod straight, a respectable distance from the scanner, looking intensely at each passenger. He only looked away to give a darting glance at the scanner screen as each bag came through. Everyone passed by timidly, grabbing their bag and scurrying toward their gate. As my bag exited the scanner, the officer gave me a hard, cold, sinister look as he jabbed his finger at me. Then, in one smooth motion, he swung his arm toward a table and emphatically pointed toward it. My world started to crumble. My mind flashed back to the words of the flight attendant on the tarmac at JFK: "Drug trafficking is a capital offense. You are responsible for everything in your possession." I was going to die in Pakistan and never get to make use of my essential oils!

In heavily accented English he pointed at my bag and spit out, "Open." First came the bottle of catsup. Feebly, I tried to explain it was a present for my son who loved catsup. Thanks to my travels, he had a collection of catsups from around the world. Next out of the backpack was the dreaded Dopp kit. He barely gave it a glance. Suddenly, I could breathe again. I wasn't going to reach the end of my life in Pakistan after all. My composure and confidence returned as I unwrapped the two bottles of Chinese apple brandy from their swaddling of dirty skivvies and placed them on the table. "What this?" he spat at me. Where my words came from I don't know. It dawned on me quickly that all he wanted was the alcohol. My response was a cool and steady, "They are presents. Can I give you one?" It worked. Like a cobra striking, he grabbed the bottle closest to him and said, "Go."

As I quickly stuffed the Dopp kit, catsup, and one remaining bottle of brandy back in my pack, I stole a quick glance at "Saddam". I could almost hear him thinking, "I just scored a big one." If he had looked my way he could have heard me thinking, "I just scored even bigger."

The flight home was uneventful. It would have been impossible to top the events of the previous three weeks. The seat on my flight home worked. I put it in recline mode, closed my eyes, and thought about a candlelight dinner and the smell and feel rhododendron oil on soft, supple skin. It was a good flight home!

MARMOT - IT'S WHAT'S FOR DINNER!

I like everything about food—from shopping for the ingredients, to preparing a meal, to sharing it with my friends. And, considering my lust for adventure in general, it should come as no surprise that I have a rather adventuresome palate. My travels have given me the opportunity to try a lot of exotic cuisines, and believe me, I'll try anything at least once. That's not to say I'll search out weird things to eat. I've never tasted brains, and I wouldn't order them off a menu. But if you order them and ask me to try them, I will. Likewise, if I ate them unknowingly and found out after the fact, I wouldn't feel the need to purge them from my system. Most likely, I'd find the meal quite pleasant, and would consider eating them again.

That said, there are a couple of foods of which I'm not very fond. One that comes to mind is trout, which I've disliked since I was a kid. That's not to say trout is bad, it's just that we ate it several times a week as I was growing up. I subsistence fished, and trout was the fish we subsisted on. The upside to that is I got good at catching dinner. The sooner I caught it, the sooner I could hop on my bike and play with my friends. The other food I have a hard time stomaching is rodent. The only rodent I've willingly eaten is squirrel. I hunt them, then donate what I catch to my friend Wayne for inclusion in his semi-sporadic Rodent Feast. Translation? The "clean all the wild game out of your freezer" meal. Wayne's chipolata squirrel is slow-cooked in a sauce of chipolata and fire-roasted peppers with diced vine-ripened Roma tomatoes, deboned, and served on a bed of rice. Hell, cooked like that, anything (even trout) would be good. I'll admit Wayne's squirrel is okay, but I'd rather have wild boar or antelope.

My limited experience with rodent grew, albeit minimally, on a trip to Mongolia. In the fall of 2012 I made an "exploratory" visit to Mongolia to search out potential fishing locations for guided tours. Through a friend of my nephew I got connected with a Mongolian cultural tour operator who was interested in expanding his business to fishing tours. Off I went to see the countryside with just a driver and an interpreter. It was here I was asked by a local family to help prepare and savor with them a traditional delicacy: fresh marmot. Now, in the U.S., we have marmots, also known as "whistle pigs", in the Rocky Mountains. They are loosely related to the woodchuck, also known as groundhogs here in the Midwest. At the time I didn't know anyone who purposely, or even accidentally, served marmot for dinner; they certainly have never been on the Rodent Feast menu. But when asked to partake, of course I agreed.

This adventure began on the plains of Mongolia, a broad expanse of land covered in short grasses that merges in places with steep hills and valleys. It's in this area where the nomadic people establish their family communities, living in portable round tents covered in animal skins known as gers (you may know this type of dwelling as a yurt). There are no roads across these vast expanses; only an occasional hint of vehicle tracks. Defining landmarks are few, yet my driver, Tsog, knew exactly where to go. As the shadows of that first evening began to lengthen we crested a small rise and three gers appeared, resting near the bank of a small river.

As Tsog pulled his battered Soviet-era cargo jeep to a halt, a woman came out of her ger to restrain her dogs and invite us in for tea and hard, dried yak cheese curds. This would be our home for a couple of days. In the plains region of Mongolia, there are no strangers, even if you are one. The welcoming nature of the people who live there comes from necessity; nomadic herders know sooner or later they will be far from their ger and need shelter for the night.

Shortly after we were settled in, I heard the high-pitched whine of a Chinese-made 125 cc dual sport motorcycle, the new horse of the Mongol hordes. The driver, the head of the family, had been out in the hills looking after the family "bank account"-- horse, sheep, goats, and yaks-- and doing a little marmot hunting. He had two marmots, each about 10 pounds, killed in the rocky edges of the hills with an old Chinese army rifle with "sticks" attached to the forearm to steady his aim. He was a good shot. Both rodents had been targeted in the head to prevent damage to any of the prime cut. With a loop around the neck just below each marmot's jaw, he hung them from a bar high enough to keep the frenzied dogs from reaching them while they aged overnight.

Apparently there is a good market for marmot. That evening, not very long after the marmot hunter returned from the hills, a local trader stopped at the ger. He wanted three kid goats for a big sack of flour, but gladly turned over the flour for two kids and a marmot.

Preparations began the next morning. Just by looking, I couldn't tell what constitutes a "prime" marmot over a "choice" marmot. But Tsog could. He spent a lot of time examining the cache, checking the thickness of the fat before deciding on the best one for dinner. In other parts of the world, meat is more of a flavoring and a treat, rather than the main ingredient.

Tsog made a circular cut through the skin where the neck joins the body. Carefully, he peeled the skin off the body down to the paws and the tail, kind of like turning a dirty sock down off your stinky feet. Once peeled, the marmots were eviscerated (the heart and liver being saved) and the carcasses cut into pieces. The marmot chunks were washed, then placed back into the skin, now turned fur-side out. Then, the future meal was stored in a tin-covered hole in the ground to keep the dogs out.

Cut into parts and pieces, an average 10 pound live-weight marmot will give a family of four plus three guests a nice feast. I am told fall is the best season for marmots, as they have developed a generous layer of subcutaneous fat to last them through the winter hibernation. The fat layer is what adds that special taste appeal. My marmot timing was perfect.

Traditionally, all marmot cooking is done by the men of the ger community. Hot rocks are placed inside the skin bag to cook the meat from the inside out. They remove the hair and crisp up the skin up by putting the skin bag directly on the fire. But my marmot was nouveau cuisine. Remember, the marmot was peeled, not skinned, to keep all that wonderful subcutaneous fat from going to waste. So what do you do with a hairy bag of marmot meat? You simply singe off the hair with a blow torch, scraping the body as you flame it, to get all the hair off. This also serves to crisp the skin and melt some of the fat to baste the meat piece inside the skin bag.

Once cooled, the skin bag is opened, the meat pieces cut into smaller units, and the skin cut into pieces. The skin and meat chunks were finished over the stove in the ger by heating rendered mutton fat using a wok-like utensil used frequently in Mongolian cooking. Next came some cut up potatoes, onions, marmot pieces, and a few hot rocks. After stirring it around for a bit, the dish was augmented with another layer of potato, onion, marmot, and rocks. Finally, a generous portion of salt and some water were added, and the wok was covered. We enjoyed some milk tea as the marmot cooked. As the marmot slowly simmered the ger took on an aroma of mutton fat and something akin to squirrel in chipotle sauce. Not bad, but not good either. For an "experience junky", the anticipation was palpable.

Marmot is a treat and delicacy. And, it was being prepared and shared with some guy from halfway around the world. We couldn't just put some marmot on our plates, find a comfortable spot on the floor, and eat it. Before the marmot could be served there were required toasts. It was an occasion. Now, I worked in Russia for a couple of years, so I learned the fine art of toasting. The elder gave the first toast, then threw back his head and poured the vodka into his mouth. Then, the next senior person toasts, and on down the line until the "honored guest" makes the final toast. The last man standing can say about anything because no one cares anymore. The Mongolian equivalent of vodka is airag, or fermented mare's milk. If you like buttermilk-flavored vodka, you would like airag.

I'm not a buttermilk fan except in making pancakes. Luckily, the women of the ger didn't participate in the toasting, so I only had to drink four shots. As I recall, my toast was something along the lines of, "Good airag, good marmot meat. Hey Buddha, let's eat." And eat we did. Marmot tastes something like a cross between rabbit and chicken. Not bad, especially after shots of airag. Being the honored guest, my plate was adorned with the choice chunks: a piece of hindquarter and a generous portion of belly skin. The hindquarter was a little stringy and the belly skin was, well, belly skin.

So how was it? I still don't care for rodent, and I'm not going to shoot one so we can have chipolata woodchuck at Wayne's next Rodent Feast! But if marmot is what's for dinner, I'll gnaw on a few of the better pieces, wash it down with some milk tea or airag if I have to, and savor the experience.

RUNNING WITH THE BULLS – LIFE LESSONS

I started traveling the world when my boys were still in grammar school. They loved to hear about my trips and couldn't wait to make one of their own. Dan got one to Scotland when he was about 12, but for some reason Ethan (E) didn't take one. I told him that when he graduated from high school I'd take him anywhere he wanted to go. For some reason we didn't go the year he graduated, but the following spring he came to me and told me he was ready for our trip—a journey to Pamplona, Spain, to run with the bulls.

Now, running with the bulls was not something I had ever thought about doing. It ranked up there with going to Norway to eat whale blubber. But when your teenage son wants to do it with you, it becomes something important. So we set off on our mission.

The most famous Running of the Bulls occurs in the Basque Country city of Pamplona, Spain—a medium-sized city of something less than 200,000 inhabitants for most of the year. The run takes place during the festival of San Fermin, which is held every year from July 6–14. Originally the "running" was a way of getting breeding bulls back from pastures and into town for bull fighting or slaughtering for a feast. Men being men, it evolved into a display of bravado by the butchers, to impress the local senioritis.

We made our way there by bus and train, traveling through Barcelona and Zaragoza. Our immediate destination after leaving the train station was the Plaza del Castillo. This iconic plaza was memorialized by Ernest Hemingway in *The Sun Also Rises*. During the festival, Pamplona's population more than triples, and it seemed all of them were in the plaza when we arrived. It was a sea of rowdy merrymakers dressed in the traditional butcher's dress of white pants, white shirt, red sash, and red bandanna. Many were drinking wine out of wine skins. E immediately wanted to get his "uniform" to be part of the crowd. But I told him, "Not yet. Anyone can just buy it. Earning it is what is important."

His retort was as I expected: "Dad, you have to make everything a life lesson, don't you?"

"Yes, E, I do. It is my job as your father."

With the matter settled, we found a table at the Café Iruña, ordered a pitcher of Sangria, and planned our run. Based on pictures from the Running of the Bulls, most people assume you race through the streets of Pamplona in front of the bulls. Nothing could be further from the truth. The bulls are released from a bullring to run for 875 meters through narrow streets. No one can outrun them. Rather, people jump into their path and run "on the horns" for a short distance before moving out of their path.

Here is how it works:

Cross streets along the course are barricaded with two fences. The first fence is right along the street, the second is about three meters back from the first. This gap gives emergency responders a place to await potential tragedy. Crowds fill the space between the fences. Once you cross the inner fence and step onto the course, you are there for the duration. If you try to climb back into the gap, the crowds, with the help of race officials, won't let you out. Trust me, I know.

The six bulls that will be fought that night are kept in a corral along with six "cut" bulls known as sweepers. The perfect run consists of three sweepers in the lead, followed by the six fighting bulls, with three more sweepers bringing up the rear. All 12 bulls are in a tight group. In a less-than-perfect run, one or two fighting bulls get separated from the group and get crazy. A bull getting separated is dangerous, but it makes for great TV coverage.

At 8 am, they fire a cannon to announce that the carrel has been opened and the bulls are on their way. When the last animal leaves the corral there is a second cannon blast. In a perfect run you hear, "Boom! Boom!."

As the bulls approach, the runners jump out in front of the bulls from along the fence, run like hell for a short distance, then duck out to the side.

Once the adrenalin subsides you go back to the Café Iruña, order at least one pitcher of Sangria, and embellish your run as you tell your story to anyone who will listen.

E and I had done our reconnaissance and chosen the starting point for our run. Our portion of the race would be less than a hundred yards from the release point. The street curved slightly to the left at the crest of a slight incline. Our rationale for picking the spot was 1) it was the first place we saw, and 2) the closer we were to the start, the less chance a bull would get separated.

At 7:30, we were inside the fences, nervously looking at our watches and waiting for our moment of glory. With each passing moment the crowd and our tension grew. At 8:00 on the dot the cannon was fired, and the corral opened. Almost immediately the second cannon blast was fired. The herd was together. At the sound of the first cannon, E jumped into the middle of the street and watched for the bulls. I, on the other hand, cautiously moved kind of toward the middle of the street. In an instant, we could hear the sound of the bulls' hooves on the cobblestone street. That sound was immediately followed by the sight of a herd of bulls, each over half a ton, charging down the street. E was off in an instant, running, as they say, on the horns. I, meanwhile, was performing a bull running move that might best be called "prancing and dancing" along the side of the herd as they thundered by. E and the bulls were gone. Suddenly, I realized only 11 bulls had passed me. Apparently one of the fighting bulls had slipped on the cobblestones and fallen, and he was charging at anything he saw moving. Lone bulls are the widow makers. As the straggler bull passed me I suddenly realized that E was not anywhere to be seen. This was back in the days before smart phones and instant connectivity.

I had no choice but to head down the cobblestone street in search for E on foot, hoping that he would not be one of the running's ugly statistics.

I finally spotted him where the street made a sharp right turn, where he was leaning against a section of fence along a side street. Not far away was a blood stain on the cobblestones. I could see that E was safe, but the blood on the street made me think about how wise a choice we had made. I would later learn E and a young woman had jumped out of the way of the herd as they approached the turn and found refuge near the fence. Neither one of them were aware that there was a lone bull coming. As they climbed the fence to avoid it, the bull brushed the woman off the fence and sent her crashing to the street. Thankfully, the blood was from a cut and not from a goring. Medics had treated her and taken her away by the time I found E.

When we saw each other, we cried out to the other, ran together, and hugged. Then, as if on cue, we both started running down the street as fast as we could go. The adrenaline overdose had to be released. We ran until we were physically and emotionally exhausted. Once we caught our breath, E looked at me with a smirk on his face and said, "Another life lesson, Dad. Let's go get your uniform, you earned it."

We did just that. Vendors were located at every street corner but we didn't make our purchase from the first one we came to. The red neckerchiefs and sashes were the same from all the vendors, but the selection of tee shirts varied. Finally, we settled on one with a big red bull emblazoned on the front. Slipping out of our running shirts we donned our hard-earned prizes and headed for Café Iruña. As we drank our second pitcher of Sangria, dressed in our bull running outfits, we turned our attention to a big screen showing a series of ESPN highlights of the day's run. One of the clips showed E climbing the fence. As I watched, I found myself thankful for two things. The first was that E's mother never saw it. The second was that they didn't catch my bull running prance and dance.

The bull running uniform life lesson was a variation on one of the life lessons I've tried to offer. It is based on lyrics from the Janis Ian song "At

Seventeen": "We all played the game, and when we dare, we cheat ourselves at Solitaire."

There are many of variations on Solitaire, but as the name implies, they are all games you play by yourself. So who is going to know if you cheat? No one, except you, so what is the big deal? You are cheating yourself, and developing a pattern of taking the easy way out of things. An earned reward is valuable. A scammed reward is not even a cheap reward. A year after our bull run, E and I went to a restaurant frequented by the local Spanish community. We wore our bull running uniforms. When we walked in, a table of older gentlemen waved us over and asked if we were "Bull Runners". When E told them we were, they ordered two bottles of the house "Sangre de Toro" to toast us as members of the fraternity of Bull Runners. As we talked, I realized I hadn't won my hand of Solitaire with the bulls. My prancing and dancing and not running was playing the Solitaire variation where you turn over all the cards. I told them about my run and received a glass of wine. If I hadn't told them, my wine, from the same bottle as E's, would have tasted bitter to me.

Being a dad, I of course have had other lessons to share. My second life lesson is about "things". I've lived in the same little starter house for more than 30 years. It's less than a thousand square feet and built on a slab with no attic. But as I think about it, the house doesn't seem that small—maybe because two rooms are only used to stash my junk and I'm often not home. If using my house as the measure, by most community standards I have not been successful. But as I look up from the keyboard and survey my domain I see a pillow from Morocco, a bowl from Russia, a carved hippo from Botswana, and a foot-high stack of journals from my travels. As I gaze at these treasures, I realize that I made a conscious choice to invest in memories rather than things. My travel souvenirs could disappear in a fire or a tornado, but at the end of my life, when they play the video highlight reel, my investments will pay big dividends.

My third lesion is about being open to life. Stuff happens to me. Sometimes it isn't what you would classify as good stuff, but most of the time, it is pretty damn good. Stuff happens because I allow it to happen.

I let life flow over and around me. I'm open to it. That doesn't mean I don't have any kind of a plan, or conversely, that I have to follow some plan exactly. It means I have this advice to give: be open to new experiences, and do as the poet Robert Frost suggests. Take the "road less traveled". Do more by accident than others will do on purpose. It will lead to some great adventures and even better stories.

Oh, and one final piece of advice-- don't play with chainsaws.

CHASING THE SUN - FOLLOW YOUR QUEST

It was so nice not to be "on stage" for a few days. Ethan and I had just deposited our most recent fly fishing clients at the Balmaceda airport in Chile and driven back to Coyhaique for a couple of well-deserved beers at our local haunt, La Pica. Over the first beer we dissected our performance as guides and our clients' skills, or lack thereof. That first beer when down very quickly. Then we ordered our second beer and dinner. Mine, of course, was pollo y papas fritas (chicken and fries), the only thing I could identify on the menu. E's was his favorite paella marine, a seafood stew.

As we sipped our second beer and enjoyed our unhurried dinner, we debated where we should fish before our next clients arrived in a few days. As we downed our third beer the sun sank lower in the sky, shining through the window and into E's eyes. As he fidgeted to keep the sun from blinding him I could see that something was hatching in his head. It was the sun's fault.

As a parent you learn very quickly how to brace yourself for your child's upcoming conversation by the way they say your name. A conversation that starts with "Daaaad" can be the harbinger of an actual or potential problem. "DAD!", on the other hand, is a sure sign of something exciting, at least for your child, even if not for you the parent. As E moved his chair out of the sun and closer to me, between spoonfuls of paella marine, out came a "DAD!"

"DAD! This is about the only place in the world where we can watch the sun rise on the Atlantic and see it set on the Pacific without flying. I bet hardly anyone has ever done it. We can do it by driving. We have three days until we have to guide again. We can fish any time, but when can do something like this again?"

I couldn't argue with his logic. It looked like an easy thing to do: cross into Argentina at Coyhaique Alto, follow highway 40 to the Pacific, sleep on the beach, watch the sunrise, cross back into Chile near Chile Chico, take the ferry across Lago General Carrera to Puerto Ibañez, and make the final dash to Puerto Chacabuco to toast ourselves with a beer and watch the sun set. It would be a simple trip of about 600 km, no sweat.

So off we set on our adventure. There was one minor issue getting across the border out of Chile, but nothing serious. We had a bit of trouble filling out the forms to take the rental car across the border, but the customs lady took pity on us and helped. Tooling along across Route 40 in Argentina, E suddenly slammed on the brakes, put the truck in reverse and said, "What the f**K was that?" He had seen a rhea, a large ostrich-like bird, back in the bush. It was time to change drivers anyway so I took over, allowing E to go on rhea watch.

Route 40 across the Argentine Pampas is a straight-as-an-arrow gravel highway. While not hard surfaced, Route 40 is an easy drive. Too easy, in fact. Before I knew it I was driving a bit faster than I should have been driving for the road conditions! Tooling along at nearly 80 km an hour I caught a whiff of something. I checked the gauges on the car—all were fine. It was the smell of rubber. As I slowed the truck and eased us to the side of the road I suddenly knew what had happened. The smell was that of rubber from a shredded rear tire. As E removed the shredded tire and I got the spare ready for him, we looked at each other and almost in unison said, "Oh shit, we don't have another spare." We had four good tires, but we were kilometers from civilization, on a gravel road that had already eaten one tire. Cautiously, we headed on down the road to the next village. The shredded tire was beyond repair and there was no tire used or new to be had. On we went, but at each new town or village we discovered the same lack of tires. All we could do was keep on, drive at a cautious pace, and hope.

We rolled into Chile Chico a little after 11 am. Instead of heading straight to the ferry launch we cruised the town looking for a place to get a spare tire, which we again didn't find, and a place for lunch. No need to hurry to the ferry launch; we were well over halfway to our goal.

After a relaxed "menu of the day" lunch of braised short ribs and potatoes we drove to the ferry launch. The ticket window was closed. Of course; it was it was lunch time. Then I looked at the sailing schedule, and realized our mistake. The ticket window wasn't closed because of lunch. It was closed because there was only one ferry a day, and it had sailed three hours earlier. Oh well, I thought; it would have been a cool thing to do, but chasing the sun wasn't my idea. We could just find a hospedaje, spend the night, and get the ferry in the morning. But E wasn't about to throw in the towel just because we hadn't checked the ferry schedule.

Out came our map of Patagonia. We calculated the kilometers between Chile Chico and Caleta Tortel, the closest location on the Pacific we could drive to—it was about 300. Checking our GPS for the time of sunset at Caleta Tortel he said, "Daaaad it will be close but we just might make it." Like a good "Magavator" (you know, the person who tells the driver they are going too fast, or should have turned left a Km ago) I checked the map and the sunset time. I wasn't sure we could make it. Almost half the distance would be right along the lakeshore, a twisty road that wouldn't allow us to drive very quickly. But this was not my quest, it was E's, so my opinion really didn't matter. I was just along for the drive. He convinced me with a final plea: "DAD, what is the worst thing that can happen. We don't make it."

"Well, E, the worst thing that can happen is, yes we don't make it, and we have another flat."

"But DAD if we don't try and wait until tomorrow we could still get another flat tire and never know if we could have made it."

"OK, let's get going," I relented. "And to paraphrase Robert Frost, we have kilometers to go before we sleep."

With E behind the wheel and my eyes alternating between the GPS and my watch, we set off on our quest. The road was even more twisted and tortuously rough along the lake shore than I had expected, but we continued to make progress toward our destination. I hoped that when we left the lake shore and reconnected with Route 7, the Pan American Highway, the road conditions would improve and we could make up the time we had lost dodging pot holes and negotiating hairpin turns along lake edge cliffs. The 45 km from our junction with Route 7 to Cochrane was dramatically better than our previous 125 km from Chile Chico. My spirits rose as E increased his speed to almost 50 km per hour as we arrived in Cochrane! We would make sunset with enough spare time to find a great viewing area and buy a beer.

Then we left Cochrane, the last "city" (about 2,500 people) in Chile. Route 7 hadn't connected Cochrane with the north and the rest of Chile until about 1990. In the ensuing 20 something years, Route 7 hadn't really pushed much further south. The good thing was we were driving a high ground clearance, four-wheel drive truck. The bad thing was we needed it, and we didn't have a spare tire. E carefully threaded his way down the "road" while I urged him to drive just a little faster. Exasperated with my "magavating", in no uncertain terms he scolded me that this was HIS quest and he wanted to make it even more than I did.

Realizing that I couldn't make the road any better or slow down the speed of the sun, I stopped checking our location or the time to sunset. With steely determination E keep his eye on the prize and drove steadily on toward the sunset. As we turned off Route 7 toward Caleta Tortel and the Pacific, I allowed myself to check the GPS. Oh my god, we might just make it. As if by magic the road conditions improved. E glanced at me and asked, "We going to make it?"

"E, we have only a few more kilometers to go and just over 15 minutes to sunset."

After what seemed like forever we suddenly arrived at the end of the road, Caleta Tortel. As E braked to a stop, he asked, "Did we make it?"

We climbed out of the truck and sat beside each other on the tailgate. With my hand proudly on his shoulder I said, "Yes, we did, and according to the GPS we have seven more minutes. "Now if only the clouds would clear and we could see it."

"Yes, that would be nice, but we did it. And you were ready to quit. Bet we are some of the few people in the world who have ever done this."

"Thanks for following your quest. Thanks to you I've done something special," I told him. "I'm so lucky to be your dad! And since this was your idea you get to buy the first round of beers."

TRAVELER OR TOURIST – THE HIMBA PEOPLE

My "victory lap" around the world was nearing its end. My son Dan had headed home, and before too much longer I would fly home from Cape Town, South Africa. But I still had a little time left, and as I sat at the hostel looking over a map, I stopped to consider my options.

I had seen most of the sights that Windhoek, Namibia, had to offer during my time as a traveler. I had been a tourist for only a couple of days during my trip. Those were the days when I floated the Mekong and visited Angkor Wat. The rest of the trip I was a traveler. To me, there is a clear distinction between touring and traveling, even if the two words are often interchanged. A tourist tends to stay in their comfort zone, sticking to major cities instead of venturing to smaller towns or off-the-beaten-path locales. They stay in areas where the amenities are similar to what they have at home. A traveler, on the other hand, explores the less-traveled areas; locations where tourism doesn't drive the economy. They interact with locals to learn and experience new things. To a traveler, a trip is a journey rather than a vacation.

I still had about three weeks left on this adventure. I knew I would see the southern part of the country when I headed toward Cape Town to fly home, so I decided to head north. As I made my plans, a young man walked over and asked if he could sit at my table. Of course he could. Fellow travelers are either a wellspring of information, or hope you will be their wellspring. In this case, we traded off between the two roles. He had rented a car in Cape Town and wanted to go north to find the Himba people. He already had one passenger, an Australian woman, but there was room for me in the car. A car trip is always better than a bus, so why not join them, I decided? The car completely packed with all our possessions, we started toward Opuwo, the center of the area inhabited by the Himba. Two days and 735 kilometers later, we arrived.

The Himba are the last semi-nomadic people of Namibia and roam over much of northwest Namibia and southwestern Angola. They are known as the "red people", receiving this name from the women's custom of grinding ocher stone (Hematite) into a fine paste, mixing it with goat butter or fat, and rubbing the mixture over their skin. The mixture gives them a rich mahogany color. The women's hair is also coated with the mixture to form long dreadlocks, except for the ends which are left uncoated and form pom-poms. The Himba are polygamists.

The Himba live in small groups or clans. A clan consists of several villages, each several kilometers from the other. Each village is led by the oldest man of the clan, basically the chief. A village consists of around 15 to 20 bee hive-shaped huts made from a frame of interlaced wooden poles plastered over with a mixture of dirt and cow dung. The huts are arranged in a circle and contained within a fence of thorn bushes woven together to form a barrier. This barrier keeps the goats and cattle in at night, and the wild world out.

During the day the men of the village are off in the bush herding their goats and cattle. The women and children – there were a lot of children since the Himba are polygamous – spend the day with domestic chores like fetching water, collecting firewood, tending small patches of maize and pumpkins, and generally keeping the village functioning.

There are a couple of Himba villages in the vicinity of Opuwo that have gone tourist. Local tour companies will drive you to one of the villages and you can spend an hour or two there watching the women dye their skin, grind maize, and make jewelry. You can even purchase their handiwork. That wasn't quite what we had in mind when we set out two days earlier. Our destination, Epupa Falls, nestled on the Angolan border 175 km from Opuwo, would take us to the heart of Himba territory. Surely we would find a "real" Himba village.

About two hours into our journey down the gravel road, I noticed what I would call, for lack of a better word, a track. It was more than just a path, but much less than a road. The track ran off at a 45 degree angle from our road. I started to watch it intently, and before long noticed what turned out to be the tops of village huts back in the bush. I got us turned around and we picked our way down the track with the car until we came to a Himba village. The Himba are known as friendly people, and they welcomed us into the village. One young boy spoke some English and happily became our guide and interpreter. Initially life in the village came to a halt as everyone looked us over and struck up conversations with us through our young interpreter. They wanted to know where we came from, where we were going, and where were our families? The villagers had open minds, and were as hungry for a new experience as we were. But before long the adults returned to their daily routine. While the children followed at a safe distance, we wandered around the village to observe daily life.

As the sun moved lower in the sky, we realized it was too late to try to reach Epupa Falls that night. I asked if we could stay in the village. Our young interpreter took us to an older woman and asked her permission, which she gave quickly. As the women of the village began preparing the evening meal we began our dinner. Ours was pasta with tomato sauce. Theirs was Mielie-meal, a porridge of stone ground corn meal cooked in soured goat's milk. Both our evening meal and that of the villagers were ready just as the men of the village returned to the compound with the herd of goats and cattle. As we shared our meals, a man in his middle to late 30's came to me with the young interpreter. Apparently, he was the chief of the clan.

As the elder of our group of three, I was assumed to be our group's chief. In most African cultures, including the Himba, a visiting chief from another clan or tribe needs to be suitably honored and shown respect. The typical showing of respect is to offer the visiting chief one of the local chief's wives. Well, given my advanced age, apparently the chief didn't think that would be the best thing to do. Instead of one of his wives, he offered me his mother! How did I get out of this? Thinking quickly, I decided to play the age card and feigned back problems. I think it worked, but I may never be the ambassador to the Himba nation. Of course, I wouldn't have had to play the age card if we had just gone to one of the tourist villages near Opuwo. But what fun would that have been.

As the sun rose, the village came back to life, and the cattle and goats were gathered to be taken out for grazing. As I rolled up out of my sleep pad and bag, the chief came over to me with a grin on his face. He reached back with is hands and patted his back. Then his grin changed into a face-wide smile and he wiggled his finger in front of our faces. My alleged back issues hadn't fooled him one bit.

MATURING STREAMS— GROWING OLDER

If I didn't know how old I am, I wouldn't know how old I am, even though my body now knows it is getting old after a long couple of days on the river. As my generation ages, the advertisers and hucksters tell us that "70 is the new 50". It sounds great, but 70 is still 70, and some physical limitations advance along with age. My mind, on the other hand, has not and will not conform to what is expected of old farts my age. My variation of what Bob Seger, the great philosopher rock star of my day said, sums up how I feel: *"He wants to live like a young man, who on more and more occasions needs a dose of Naproxen, with the wisdom of an old man."*

On November 12, 1990, at the base of Mount Cook in New Zealand, the concept of getting older first became real to me. I met a young couple at the hostel at Mount Cook National Park. They were going to climb the mountain, cross the Copeland Pass, and hike down into Westland National Park. It sounded like a fun adventure, but I couldn't join them. Ice axes and crampons were needed to make it over the pass, and I didn't have the gear. Besides, it was a long and arduous climb, and the weather was too unpredictable to make it in a single day's assault. However, a short day's climb ended at a staggeringly picturesque location, Hooker Hut. After a night there, with an early start, these climbers planned to scurry over the pass and make it to another hut on the Westland side before nightfall, then traverse the final leg on the following day. Mount Cook called to me, too, and as I stared at the summit I felt like a pre-teen school boy with a crush on the most alluring girl in school, the one you want so much to be your square dance partner, but are just as afraid that she might say yes as you are of being rejected. Funny how a mountain can do that to a grown man. I wished I had the gear to join them, even if I had secret doubts about my ability to do it. But given their invitation, I couldn't resist agreeing to help them carry their gear to the Hooker Hut and getting a closer feel for their journey and this powerfully seductive mountain.

Just after breakfast, as I waited in front of the crowded Aoraki Alpine Lodge for my new acquaintances, a tour group came out to begin boarding their tour bus. One man, who back then I would have described as an "older gentleman" but now would describe as "like me", looked at my backpack.

He asked, "Are you going to climb that mountain?"

I told him, "No, I am just going up as far as the staging hut."

With a sigh and a slight shake of his head he said, "Why did I wait until now to see the world instead of starting when I was young like you?"

Ironically, I had had the exact same thought about traversing the pass as I waited for my younger companions. When I was their age, I was busy with the Marine Corps, a wife, and going to college. I have no regrets, but I'm not immune to a few would haves, could haves, and should haves now and then. Now, as I approach my seventh decade, I often jokingly say to my similarly-aged companions that if I get going now I won't die with as many "didn'ts", or "maybe can'ts", but rather with a few more "dids"!

I remember thinking about my brief exchange with that old tour bus rider a good deal during the remainder of that trip. Once I was home, however, the feeling faded away as I settled into my typical routine. Hell, I was just 43 then. I was in the mid reach of life—not as much raw plunging energy, but still, there were periods of power and not much reflection. These days, though, I am flowing through the lower reaches with a slower pace and with greater depth and breadth. Let me explain.

In a previous life, I was a fisheries biologist and a water resources engineer. A guiding principle in that endeavor was something called the "River Continuum Concept." The phrase is an appropriate metaphor for the aging process. Basically, it's a system for classifying how a stream changes and its relationship of equilibrium with the landscape as it flows from its headwaters to the ocean. A river's width, depth, gradient, and temperature change constantly as it flows toward the ocean, as does its immediate environment—not unlike our lives from birth to death.

Hameshop Brook, the brook that flows past where I grew up in New Hampshire, begins high up on the side of Mt. Kearsarge. It's named for the old Hameshop, a turn-of-the-century factory that made harness collars for horses. The Hameshop closed well before I was born. By the 1930s there was no longer a big demand for horse collars. During the war, however, it re-opened for a short time to can water from the brook for the war effort.

Near its source, the brook flows through a forest of hemlocks, white pines, and the occasional beech trees that fight to hold out against the rocky soil. The gradient of the brook up high is steep, with a series of rapids, plunge pools, and a waterfall or two; this section is known as the headwaters. Like most mountain streams, for a short stretch it becomes wider and slower, perhaps because of a beaver pond or some quirk of geology. For Hameshop Brook, this brief respite is Bradley Lake, which now boasts about 180 acres. Many, many years ago, most likely to power a mill, a dam was built to raise the level of the lake. Mills and dams were almost always built to take advantage of a change in a stream's gradient and power. Bradley, as we called it, became two lakes separated by a narrow channel with a high point, Blueberry Island, in the middle.

Hameshop Brook leaves the lake with high energy again through a narrow but steep overflow on the northeastern side. Here, it's not as wild as it was upstream of the lake, but it's still full of erosional, mostly downwardly directed, energy. As a boy growing up along Hameshop Brook, I was full of this type of energy. I thought I could do anything, and tried to. As counter-intuitive as it may sound, the Bradley Lake of my life was the Marine Corps—the lake in this analogy. There, my energies changed from carving my channel to storing some potential energy. The stored potential energy was used to get me through college at Eastern

Kentucky University with degrees in fisheries biology, and the University of Missouri with a degree in water resources engineering. With still some energy left, I applied it to starting my career and, unknown to me, impelling the creation of the next phase of my life.

Someplace just below the long-abandoned old Hameshop, the brook's character changes as it enters what is known as the mid reaches. The mid reaches of streams consist predominantly of pools. Much of the erosional energy is still directed downward, but more of it is directed toward the banks. The water is deeper, the gradient is much less steep, and the flow is slower than in the Headwaters. For me, the early midreaches of life, my late 30s and early 40s, corresponded to the development of my career and a complete change in my life. I'm not sure if the critical change point in my life was the last rapid of the headwater reach or the first riffle of the mid reaches.

In the midreaches of streams, an interesting phenomenon occurs. Approximately every five to seven times the average width of the stream there is a short rapid or riffle. The stream becomes a series of long pools separated by shorter, shallower, fast-moving water. My career was the midreaches pools of my life. I had found my niche doing water quality studies, analyzing big water quality data sets, and publishing the results. As in the stream, most of my 40s and 50s were spent in the pool environ. I moved steadily along like the water through the pools. You don't see the erosion in the pools, but it is there. But every so often, inevitably and with some regularity, my pace increased as life hit a riffle. Riffles can consist of troubles—relationships, money, tragedies. But they can also be high points—releasing my pent-up gene for wanderlust, meeting someone special, being a dad. Riffles between the stream's pools create renewed energy. In life, these riffles release our stored energy, and if we let them, remember and try to recapture the headwaters of childhood.

To the Hameshop Brook, the ocean is the Blackwater River. Like the transition from headwaters to midreaches, there is no clear point of demarcation between midreaches and lower or meandering reaches. For Hameshop Brook, the transition to the lower reaches begins when the trees shading the brook begin to thin out and eventually disappear altogether. In aging, this transition resembles losing our hair—eventually, the old brook's banks are bald. In river terminology, this is called the meandering stage, and nearly all of the erosional energy is directed toward the sides and not downward. The Hameshop Brook erodes its banks as it meanders slowly across the meadow. In middle age, the 50s to early 60s, it's time to expand and get comfortable, pay off houses, prepare for retirement, and rest on a reputation built in the earlier midreaches. In the meadow, there are no riffles. For me, the midreaches of middle age held the conclusion of a steady and interesting career that transitioned to lower reaches in the form of early retirement, and time to lazily view life's options and explore new opportunities. Lateral erosion is not dramatic like the downward erosion of the headwaters or even the riffles in the midreaches, but it's powerful in a different way. Where the sharp bends of the meanders scour and undercut the banks, big trout hide and memories and dreams find shelter

Sometimes in these lower reaches, two bends might actually erode into each other, forming an oxbow lake. The isolated section then slowly fills in, and the water within it can never make its way to the ocean. When there is no renewing water from the flow, even the slow waters of this meadow stage, there is no erosion. The oxbow is stagnant and slowly fills over time. Sometimes something completely different can happen. The stream forms one or more channels as it flows across its valley. Side channels or braids occur when something obstructs the stream's flow. The stream takes the easy way, and rather than exert all its energy in eroding away the obstruction, just erodes away enough of the meadow for some of the flow to bypass the impediment. After their own meanders, as quickly as conditions allow, the braid rejoins the main flow and the journey to the ocean. The meanders and braids of our lives reflect the changes in the composition of the banks, bed, and obstructions to our flow. Instinctively knowing, because of how our velocity has slowed, that we are approaching our ocean, we can take our remaining precious time to explore as much of the valley as our flow allows, and enjoy a variety of potential paths and channels that the meadow has to offer before we reach the ocean.

There is a seasonal phenomenon associated with stream flows. In the fall, as the trees in the watershed lose their leaves, the base flow of the stream increases just a bit. As winter approaches and their leaves fall, the trees don't need as much water. No longer making sugars, they take up less ground water and transpire even less into the atmosphere. Suddenly there is more water in the stream. For a short period, before the ground freezes and the groundwater becomes locked in the soil, the stream's flow increases. This is reflected in a slight, but still perceptible increase in the stream level and the erosional power. This autumn increase in erosional power occurs in all the stream's reaches but is most pertinent in the lower reaches. In my early autumn I'm beginning to sense that the chlorophyll is being pulled back from the leaves and the yellow and golden pigments are becoming more evident. I know how I'll use that late season burst of energy. I'll go adventuring. When the ground finally freezes I'll sail off on an ocean adventure.

So in reply to the old man's lament at the tour bus at Mount Cook, according to the River Continuum Concept, coulda shoulda woulda is a moot point. Now having just turned 70, I can say this about life: rivers don't have the same energy, either in direction or intensity, as they become more mature. It's not really a question of whether the young stream got far enough out of its banks or tumbled over enough waterfalls, or whether the meanders through the meadow have wasted time. It's certainly not a question of whether the lower reaches can fly high again. It's a question of how we collect the moments as we travel down to the ocean—how we remember and consider what we are doing when we are doing it rather than looking back with perhaps sadness, but not regrets.

Made in the USA
Lexington, KY
06 March 2018